SpringerBriefs in Computer Science

More information about this series at http://www.springer.com/series/10028

Atle Refsdal · Bjørnar Solhaug
Ketil Stølen

Cyber-Risk Management

Springer

Atle Refsdal
SINTEF ICT
Oslo
Norway

Ketil Stølen
SINTEF ICT
Oslo
Norway

Bjørnar Solhaug
SINTEF ICT
Oslo
Norway

ISSN 2191-5768   ISSN 2191-5776 (electronic)
SpringerBriefs in Computer Science
ISBN 978-3-319-23569-1   ISBN 978-3-319-23570-7   (eBook)
DOI 10.1007/978-3-319-23570-7

Library of Congress Control Number: 2015950450

Springer Cham Heidelberg New York Dordrecht London

Printed on acid-free paper

Springer International Publishing AG Switzerland is part of Springer Science+Business Media
(www.springer.com)

Preface

Information and communication technologies (ICT) have over several decades brought significant benefits to enterprises, individuals, and society as a whole. This is clearly evident when considering the wide and profound impact of the Internet in a great many parts of our daily lives. The Internet, and more broadly cyberspace, has become a cornerstone for a broad range of services and activities that today we take for granted. Due to cyberspace and its underlying infrastructure, people and organizations have access to more and better services than ever before. This is the case within several domains of society, including banking and finance, communication, entertainment, health, power supply, social interactions, transportation, trade, and social participation. As a result, our daily lives, fundamental rights, economies, and social security depend on ICT working seamlessly.

At the same time, cyberspace has introduced, and continues to introduce, numerous new threats and vulnerabilities. Stakeholders are exposed to cybersecurity incidents of many different kinds and degrees of severity. These include information theft, disruption of services, privacy and identity abuse, fraud, espionage, and sabotage. At a larger scale, societies are threatened by possible attacks on critical infrastructures via cyberspace, as well as the potential for cyber-terrorism and even cyber-warfare. In addition to the many possibilities for cyber-crime and malicious attacks come all the accidental and other non-malicious threats that may lead to cybersecurity incidents. In fact, the ubiquity of cyberspace has brought societies to a point where a very large number of the risks that we traditionally have been exposed to in the physical world today arise in cyberspace and have become cyber-risks.

In order to ensure a satisfactory level of cybersecurity, stakeholders need to understand the nature of cyber-risk and what distinguishes cyber-risk from other kinds of risk, and they need adequate methods and techniques for cyber-risk management. Our main objective with this book is to give a short introduction to risk management, focusing on cybersecurity and cyber-risk assessment. We introduce the reader to the underlying terminology, we present and explain the processes of cyber-risk management, and we provide guidance and hands-on examples on how to conduct cyber-risk assessment in practice. We moreover address many of the typical challenges that risk assessors face, and we give advice on how to tackle them.

There are many different techniques, tools, modeling languages, and documentation formats that are available to support cyber-risk assessment. This book is obliv-
ious to any such specific approach; while we have based the contents on established
standards and industry best practices, we present the risk assessment process and
the examples in a format that can be instantiated by any specific approach that com-
plies with the ISO 31000 risk management standard. The intended target audience
is practitioners, as well as graduate and undergraduate students, in particular within
the ICT domain. We also aim to provide lecturers with teaching material on the
fundamentals of cyber-risk management and the basic principles and techniques of
cyber-risk assessment. We moreover believe that the book illuminates and clarifies
many aspects and underlying concepts of the domain of cybersecurity. The book
can therefore be useful also for researchers and standardization bodies that have
activities related to cybersecurity.

Our own knowledge about and experience of cybersecurity and cyber-risk management, and therefore also the contents of this book, largely stem from academic
research and empirical studies that we have conducted jointly with colleagues and
with collaborators from industry. We express our acknowledgments to all of those
who in different ways have helped out in the work on this book.

We owe many thanks to our close colleagues Gencer Erdogan, Yan Li, Aida
Omerovic, and Fredrik Seehusen for their many and valuable comments and sug-
gestions on several parts of this book. We are very grateful to Kristian Beckers,
Karin Bernsmed, Aslak Wegner Eide, Marika Lüders, and Ragnhild Kobro Runde
for reviewing the manuscript and providing good and helpful feedback.

Prior to and during the work on this book we have benefited greatly from col-
laboration with people from academia and industry on several research projects.
These include Jürgen Großmann, Maritta Heisel, Fabio Martinelli, Wolter Pieters,
Alexander Pretschner, Christian W. Probst, and Aristotelis Tzafalias.

Some of the research activities that the work on this book has benefited from have
partly been funded by the Research Council of Norway, in particular through the
projects Diamonds and AGRA. Relevant research activities have also been funded
by the European Commission, in particular through the projects RASEN and NES-
SOS, but also through CONCERTO.

Oslo, Norway   *Atle Refsdal*
July 2015   *Bjørnar Solhaug*
                                                                                Ketil Stølen

Contents

Acronyms

AMI	Advanced metering infrastructure
ARPANET	Advanced Research Projects Agency Network
CAPEC	Common Attack Pattern Enumeration and Classification
CIIP	Critical information infrastructure protection
CIP	Critical infrastructure protection
CNSS	Committee on National Security Systems
COBIT	Control Objectives for Information and Related Technology
CWE	Common Weakness Enumeration
DDoS	Distributed denial of service
DoS	Denial of service
ENISA	European Union Agency for Network and Information Security
EU	European Union
EUROPOL	European Police Office
GAO	US Government Accountability Office
GPRS	General packet radio service
ICT	Information and communication technology
IEC	International Electrotechnical Commission
ISACA	Information Systems Audit and Control Association
ISO	International Organization for Standardization
IT	Information technology
ITU	International Telecommunication Union
LAN	Local area network
NIACAP	National Information Assurance Certification and Accreditation Process
NIST	National Institute of Standards and Technology
NSFNET	National Science Foundation Network
OWASP	Open Web Application Security Project
SLA	Service level agreement
UML	Unified Modeling Language
WAN	Wide area network

Chapter 1
Introduction

The cybersecurity strategy of the European Union (EU) [13] begins by stating that over the last two decades the Internet, and more broadly cyberspace, has had a tremendous impact on all parts of society. Our daily life, fundamental rights, social interactions, and economies depend on information and communication technology (ICT) working seamlessly. The US strategy for cyberspace [83] argues accordingly and quotes US President Barack Obama: "This world – cyberspace – is a world that we depend on every single day … [it] has made us more interconnected than at any time in human history."

On the other hand, as argued by the abovementioned strategy of the EU, while the digital world brings enormous benefits, it is also vulnerable. Cybersecurity incidents, be they intentional or accidental, are increasing at an alarming pace and could disrupt the supply of essential services we take for granted, such as water supply, health care, electricity supply, or mobile services. In the 2015 report on global risks [87] the World Economic Forum ranks "cyber attacks" among the top ten risks in terms of likelihood while "critical information infrastructure breakdown" is among the top ten risks in terms of impact.

The cybersecurity strategies of the EU and nations worldwide aim at making sure that players in a number of key areas (such as energy, transport, banking, stock exchanges, and enablers of key Internet services, as well as public administrations) assess the cybersecurity risks they face, ensure that networks and information systems are reliable and resilient via appropriate risk management, and share the identified information with the competent national authorities.

Moreover, as witnessed by a number of annual reports on cybersecurity, cyber-risk has become a persistent, all-encompassing business risk for a great many enterprises and organizations. According to the PwC Global State of Information Security Survey 2015, for example, cyber-risk "is no longer an issue that concerns only information technology and security professionals," and "incidents and financial impacts continue to soar. [66]"

In order to correctly assess cybersecurity risks there is a need for reliable methods, tools, and processes for risk management in general, and for cybersecurity risk assessment in particular. Moreover, these methods, tools, and processes must be

© The Author(s) 2015
A. Refsdal et al., *Cyber-Risk Management*, SpringerBriefs
in Computer Science, DOI 10.1007/978-3-319-23570-7_1

made available to their intended users through special-purpose courses and educational materials. This short book in the form of a technical brief on cyber-risk management is meant as a contribution in that direction.

The remainder of this chapter is divided into five sections. We start by presenting the aim and emphasis of this book followed by our policy of writing and presentation. We then go through the overall structure of the book and the decomposition of its three main parts. Thereafter we specify our intended readers and some ways to read the book. Finally, we provide an overview of relevant standards and how we have used them.

1.1 Aim and Emphasis

Our overall objective with this book is to give a short and focused introduction to risk management, with particular emphasis on cybersecurity and cyber-risk assessment, building on best practices from industry.

This book builds on and is complementary to established standards in several respects. First, the book defines and explains the background of the terminology to give a more thorough understanding of the domain of cyber-risk management. Second, the book has a pragmatic orientation in that it explains not only what cyber-risk management is (as the standards do), but also how to do it. Third, the book gives a running example that illustrates the various tasks of cyber-risk assessment and how to conduct them. Fourth, the book addresses several of the typical challenges that assessors of cyber-risk encounter, and provides advice on how to tackle them.

We have tried to write a book that distinguishes itself from other books on the same topics in that it serves as a brief and general introduction to cybersecurity and cyber-risk management, without being related to specific existing approaches and techniques. At the same time we focus on the pragmatics, addressing questions such as: How should cyber-risk assessment be conducted? Which techniques should be used when? what are the typical challenges and problems that may arise? How should they be handled?

1.2 Policy of Writing and Presentation

This book is written in the "we-form." However, we use "we" with a special risk assessor interpretation. From Part I onwards, "we" refers to the assessment team consisting of the assessment leader, the assessment secretary, and the reader of this book. We think of the latter as a trainee in risk assessment; "you" is used to refer directly to the reader who we assume plays the role of a trainee.

As already mentioned, with respect to methodology for risk management and assessment, we have written the book at a generic level without favoring any specific approach. The risk scenarios are documented using a straightforward and generic table format.

In the case of definitions, we use a specialized environment embedding each definition in a gray box. Moreover, we use UML class diagrams [58] and Venn diagrams [9] to summarize and relate the various definitions graphically.

1.3 Structure and Organization

This book gives a general introduction to the central concepts and notions of risk management. The book provides an overview of cybersecurity and cyber-risk assessment, the involved tasks, and how to conduct them in practice. We present and discuss the main challenges that practitioners typically encounter, and we offer advice on how they should be handled. The book starts with a preamble consisting of a preface, a list of acronyms, as well as the introductory chapter that you are currently reading. The preamble is followed by three main parts as further detailed below, and ends with an addendum consisting of a concluding chapter followed by a glossary, a bibliography, and an index.

1.3.1 Part I: Conceptual Introduction

Part I gives a conceptual introduction to the topic of risk management in general and to cybersecurity and cyber-risk management in particular. The focus is on cyber-systems and cybersecurity, and the introduced terminology builds on established international standards. Part I is divided into four chapters:

- *Risk Management* – This chapter gives a brief introduction to risk, risk management, and risk assessment in general.
- *Cyber-systems* – This chapter introduces the notion of cyber-system. What are the characteristics of cyber-systems and how are they related to cyberspace?
- *Cybersecurity* – This chapter introduces the notion of cybersecurity and describes its connection to the related concepts of information security, critical infrastructure protection, and safety.
- *Cyber-risk Management* – This chapter introduces the notion of cyber-risk and specializes the general process of risk management to cope with cyber-systems, cybersecurity, and cyber-risk.

1.3.2 Part II: Cyber-risk Assessment Exemplified

Part II of the book presents the main stages of cyber-risk assessment from context establishment to risk treatment. A running example is used for illustration purposes. The example concerns an advanced metering infrastructure (AMI) in a smart grid.

The example and results presented have been made up in order to provide a demonstration of the assessment process that is simple to understand also for readers not familiar with (or interested in) smart grids. All descriptions are at a generic level, and we do not go into technical details. The material we present does not pertain to any real system.

Part II is divided into five chapters which together exemplify each step of a cyber-risk assessment, from context establishment to risk treatment:

- *Context Establishment* – Establishing the context involves setting the risk evaluation criteria and purpose of the cyber-risk assessment. It also involves defining the scope and focus of the assessment, and describing the target of the assessment.
- *Risk Identification* – Risk identification involves determining what could happen to cause potential harm to assets (the valuables we aim to protect). This includes gaining insight into how, where, and why such incidents may occur, irrespective of whether the source of the cyber-risk is under the control of the party on whose behalf the cyber-risk assessment is carried out.
- *Risk Analysis* – Risk analysis involves determining the level of cyber-risk, typically in terms of the likelihood of incidents to happen and the consequence for assets. This can be done qualitatively or quantitatively.
- *Risk Evaluation* – Risk evaluation is the task of comparing the results of the risk analysis with the risk evaluation criteria (defined during context establishment) to determine whether the cyber-risks need treatment. It also involves aggregation and grouping of risk that should be considered together.
- *Risk Treatment* – Risk treatment involves deciding on strategies and controls to deal with cyber-risks. It also involves deciding to accept the (residual) cyber-risk, and formally recording the decisions and responsibilities.

1.3.3 Part III: Known Challenges

Cyber-risk management and cyber-risk assessment involve many challenges, and some are more difficult than others. In Part III of the book we focus on four of the most important ones. We explain what each challenge consists of and under what conditions it appears. We also offer recommendations on how each challenge can be handled from a practical point of view. The challenges addressed are:

- *Which Measure of Risk Level to Use?* – There is no universal agreement on how to measure risk. The definition of risk in ISO 31000 [25], for example, comes with five notes, each defining risk in a slightly different way. Traditionally, risk value is a function of two factors, namely likelihood and consequence. However, within the field of cybersecurity three-factor and many-factor definitions are gaining popularity. This chapter discusses the different alternatives and provides advice on when to use which.

- *What Scales Are Best Suited Under What Conditions?* – The selection of the right scale for the right purpose is essential. The selection of scales is particularly important when measuring expert judgments. This chapter gives an overview of relevant kinds of scales and provides advice on which to use when and how the scale should be defined. The chapter also discusses the strengths and weaknesses of qualitative versus quantitative scales. When should we use which, and does it make sense to combine?
- *How to Deal with Uncertainty?* – In relation to risk assessment the issue of uncertainty appears at several levels. We may talk about uncertainty in the meaning of a specific risk appearing with some likelihood. We may also talk about how certain we are that this estimate of likelihood is correct. In the latter case, we basically estimate our trust in the former estimate. In this chapter we give recommendations on how to handle the various forms of uncertainty in practice.
- *High-consequence Risks with Low Likelihood* – Risk assessment is said to be unreliable for risks of low likelihood and very high consequence. In this chapter we explain why, and offer guidelines on how to deal with such situations. We also discuss the problem of the "unknown unknown", often referred to as the "black swan problem."

1.4 Intended Readers and Ways to Read

Our intended target audience is practitioners and (under-)graduates who are interested in the fundamentals of cyber-risk management and the basic principles and techniques of cyber-risk assessment, as well as lecturers who need teaching material on these topics.

Depending on the background and interests of the reader, there are many ways to read this book. Readers with little background in the field might read the chapters in sequential order. A more experienced reader will perhaps only look up certain sections that he or she finds particularly interesting. Nevertheless, we would like to provide some advice:

- Chapter 2 introduces risk management in general. Readers who are already acquainted with risk management may skip that.
- Part I is sufficient to get an overview of cyber-risk management.
- Part I and II are sufficient to also get an overview of cyber-risk assessment.
- Part III is more specialized and mainly relevant for readers who want to conduct cyber-risk assessments themselves.

1.5 Relevant Standards

There are many relevant standards for cyber-risk management. ISO 31000 [25] provides principles and generic guidelines on risk management. The standard is not specific to any industry or sector. It can be applied throughout the life of an organization, and to a wide range of activities, including strategies and decisions, operations, processes, functions, projects, products, services, and assets. It can be applied to any type of risk, whatever its nature, whether having positive or negative consequences.

ISO/IEC 27005 [32] provides guidelines for information security risk management. These guidelines are based on ISO 31000. ISO/IEC 27005 supports the general concepts in ISO/IEC 27001 [33] on requirements for information security management systems. It is designed to assist the satisfactory implementation of information security based on a risk management approach. ISO/IEC 27032 [28] offers guidelines for cybersecurity with particular focus on the virtual world and virtual, intangible assets.

Many specialized or domain-specific standards are based on ISO 31000 and ISO/IEC 27005, while others offer guidelines or mappings characterizing their relationship to these. An example in this respect is COBIT [38] which has been developed by ISACA. COBIT is a framework for information technology (IT) management and IT governance. Its relationship to ISO/IEC 27005 is described in "COBIT for Risk" [39]. Similarly, ITU-T X.1055 [36] on risk management maps ISO/IEC 27005 to the telecommunication domain. Kristian Beckers [3] has explored the relations between security standards and methods for security requirements engineering.

In this book we do not restrict ourselves to any standard in particular. What we put forward, however, fits well within the general framework of ISO 31000 and ISO/IEC 27005.

Part I
Conceptual Introduction

Chapter 2
Risk Management

The topic of this chapter is risk management in general. We begin by explaining what risk is and presenting the terminology we need in order to talk about risk. Thereafter we introduce risk management and explain what it involves for an organization to manage risk in a systematic and effective manner. Subsequently we look more into the details of the risk management process and its sub-processes.

2.1 What is Risk?

Basically, risk is the potential that something goes wrong and thereby causes harm or loss. The gravity of a risk depends on its likelihood to occur and its consequence. The consequence is the impact on an asset, and an asset is an object of value that we want to protect.

> **Definition 2.1** A *risk* is the likelihood of an incident and its consequence for an asset.

In order to convey more precisely what this definition means, we need to explain the concepts it refers to, namely incident, likelihood, consequence, and asset. We start with the notion of incident. When we discuss or assess risk we need to be careful to distinguish between its causes and the potential occurrence of the incident that constitutes the risk. Consider, for example, a burglar who enters a house by breaking in through a window. Understanding how such events unfold and what makes them possible is necessary for understanding how risk arises. But the actual incident itself is only the event that causes the harm or loss. In our example such an event could be the theft of jewelry. The definition of incident makes this precise.

> **Definition 2.2** An *incident* is an event that harms or reduces the value of an asset.

The definition of the term "incident" makes it clear that risk is about the occurrence of harmful events. But when can we say that an event is harmful and therefore an incident? After all, this depends on who we ask and what our focus is. For example,

© The Author(s) 2015
A. Refsdal et al., *Cyber-Risk Management*, SpringerBriefs
in Computer Science, DOI 10.1007/978-3-319-23570-7_2

could we not say that a burglar breaking a window lock is a harmful event? And
could we not say that a burglar entering a house is at the cost of privacy? The an-
swer to these questions is that it depends on what our assets are. If our concern is
the window lock, then the breaking of the lock is an incident. On the other hand,
breaking a window lock alone does not harm privacy, and is therefore not an incident of a privacy risk. By including the notion of asset in the definition of incident
we are forced to be specific about which events can be understood as incidents.

Definition 2.3 An *asset* is anything of value to a party.

The party is the entity or unit, such as a company or other organization, for which
the assets in question have value. In the same way as there is no risk without an asset,
there is no asset without a party. Because what is held as assets and how valuable
they are depend on the party, we always need to be specific about who the party is
when we manage or assess risk.

Definition 2.4 A *party* is an organization, company, person, group, or other body
on whose behalf a risk assessment is conducted.

Hence, before we can discuss or assess risk, we must determine the party and the
assets of concern to this party. Only then can we speak of incidents and risks in a
precise and meaningful manner. In the burglar scenario, for example, the party could
be the house owner. But it could also be someone renting the house. Whereas the
assets for the renter may include both jewelry and privacy, the house owner would
typically worry about damage to the property.

Notice that a party is not the same as a stakeholder. A party may be thought of
as a stakeholder, but in a risk assessment situation there are normally many stakeholders that are not parties. A *stakeholder* in this context is basically any person or
organization that may affect or be affected by the subject of the assessment. If we
conduct a risk assessment on behalf of a company then the company is the party.
Within and related to the company there may be many stakeholders (for instance,
employees and suppliers) with all kinds of conflicting interests and they are not parties in this risk assessment. When we identify assets on behalf of a party we focus
solely on the interests of the party in question. In most risk assessments there is
only one party. If, however, there are several parties then the assets of the different
parties must be kept apart. The same object, for example, a human life, may be an
asset of different values for different parties. For you, the value of your life is perhaps infinite, while for a hospital it may be equal to the amount they have to pay the
bereaved in the case of death due to maltreatment by the hospital.

The remaining concepts from the definition of risk are those of likelihood and
consequence. Together, these notions characterize the gravity of a risk.

Definition 2.5 A *likelihood* is the chance of something to occur.

The notion of likelihood refers to the chance of something happening, no matter
how we measure or represent it. Sometimes we describe it qualitatively, and other
times quantitatively. We may describe it in general terms, and we may represent it
mathematically as probability or frequency. *Probability* is a measure of the chance of

occurrence expressed as a number between 0 and 1, whereas *frequency* is a measure of the number of occurrences per unit of time.

> **Definition 2.6** A *consequence* is the impact of an incident on an asset in terms of harm or reduced asset value.

In this book the term "consequence" refers to negative impact only. Some approaches to risk management take a more general view of risk by considering any effect on assets, both positive and negative. This is useful when we conduct risk management with the aim of balancing risk and opportunity; negative outcomes could be accepted given the foreseen gain.

As mentioned above, the notions of likelihood and consequence characterize the gravity of a risk. We measure this gravity in terms of risk level.

> **Definition 2.7** *Risk level* is the magnitude of a risk as derived from its likelihood and consequence.

Our definitions of risk and risk level are well established and widely used. There are, however, alternative ways of expressing and measuring risk, some of which we discuss in Chap. 11.

The UML class diagram of Fig. 2.1 illustrates how the terms we have defined in this section relate to each other. Risk consists of three ingredients, namely consequence, incident, and likelihood. The relation represented by a line with a black diamond connecting risk and consequence captures that consequence is an ingredient that belongs to risk. The consequence represents the impact of an incident on an asset. Consequence is therefore also connected to the relation between incident and asset, since it is a measure of harm. The diagram also captures that for a given asset, there is a party that values it.

The same incident may give rise to several risks. Risk is therefore connected to incident with a white diamond to express that although incident is an ingredient of risk, it does not necessarily belong uniquely to one risk. Likelihood is a measure of how often the incident occurs. We may therefore see likelihood as an ingredient that belongs to incident. Since incident is an ingredient of risk we also have that likelihood is an ingredient of risk.

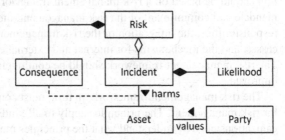

Fig. 2.1 Risk concepts

The terminology we just introduced allows us to explain what risk *is*, but not how risk *arises*. In the rest of this chapter we look more closely into how risks are managed, which includes understanding the sources and causes of risk and how to handle them.

2.2 What is Risk Management?

All organizations are exposed to risk, and most organizations do some kind of risk management. However, if we aim to precisely understand the kinds and nature of the risks, and to manage them in a systematic and effective manner, we need a well-defined process for risk management. We moreover need to understand the underlying principles and framework for the risk management process.

Definition 2.8 *Risk management* comprises coordinated activities to direct and control an organization with regard to risk.

For a risk management process to be adequate, efficient, and effective it should be based on a risk management framework. This framework should in turn comply with the basic principles for risk management. These relationships between the risk management principles, framework, and process are shown in Fig. 2.2. The framework should be subject to continual improvement, partly based on experience, findings, and results from the risk management process. This explains the arrow back from the process to the framework in Fig. 2.2. The purposes of the risk management

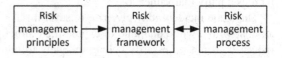

Fig. 2.2 Risk management elements

process must be decided as part of the overall management of the organization. This is why the implementation of the risk management process in the organization should be based on a risk management framework. The framework defines the mandate and commitment of the risk management, the risk management policy and responsibilities, the integration of the risk management into the organizational processes, and the mechanisms for internal and external communication and reporting. The risk management framework should be continuously monitored, reviewed, and improved.

The risk management framework, in turn, must comply with the basic principles for risk management. The principles apply to all kinds of risk management, but organizations need to understand what the principles mean for them and for their own framework for risk management. ISO 31000 lists eleven such principles. Among others, these include the principles that risk management shall create and protect value, that risk management shall be an integral part of all organizational processes, that risk management shall be part of decision making, and that risk management shall be based on the best available information.

Figure 2.3 presents the risk management process in more detail. *Risk assessment* is a finite process that organizations conduct on a regular basis. The two others, namely *communication and consultation* and *monitoring and review*, are continuous activities. In the following we look more closely into the details of these three components of the risk management process.

Fig. 2.3 Risk management process

2.3 Communication and Consultation

By *communication and consultation* we mean activities aiming to provide, share, or obtain information and to interact with stakeholders regarding the management of risk. A *stakeholder* in a risk management context is a person or organization that may affect or be affected by the organization that is the subject of the risk management.

The interaction and information sharing serve as a basis for decision making. The information of relevance is anything that may determine how the organization should manage risk, including how risks should be communicated to internal and external stakeholders. This includes both external issues such as legislation, market situation, and external sources of risk, and internal issues such as reorganization, business strategies, and risk appetite. Such information can in general relate to the existence, nature, form, likelihood, significance, evaluation, acceptability, and treatment of risk [25].

For the communication and consultation to be efficient and effective it is advisable to establish a dedicated team and to define a plan for the process. This, in turn, helps to ensure endorsement of the risk management process and to communicate risk assessment results as explained in the following.

2.3.1 Establish a Consultative Team

The communication and consultation with internal and external stakeholders may concern any part or activity of the overall risk management process. Efficient and adequate communication and consultation ensures that those responsible for implementing the risk management process understand the basis for decisions and why particular actions are required. As part of the overall risk management it is useful to establish a consultative team with defined responsibilities for the communication and consultation. Such a team typically includes internal stakeholders such as decision makers and risk managers, as well as employees with insight into the organization. The team may also include external stakeholders such as board members, customers, and those with a vested interest. The roles and responsibilities of the team members must be clearly defined and specified. For smaller organizations it may not be an option to establish a team for the communication and consultation. In that case, the organization should still appoint a responsible point of contact and consultation.

2.3.2 Define a Plan for Communication and Consultation

The way risks are judged and perceived varies from person to person, even within the same organization. This may be due to differences in background, position, values, needs, concerns, and so forth. Decision makers need to take such varying perceptions into account when determining how to manage risks. A clear plan and good procedures for communication and consultation aid decision makers in this respect. The consultative team, or those responsible for the communication and consultation, should be involved in defining the plan and procedures. In addition to defining roles and responsibilities, organizations should establish procedures for how to support any of the processes of the overall risk management. This includes, for example, ensuring that different areas of expertise are brought together during risk assessments, that the interests of all relevant stakeholders are considered, that the risk evaluation criteria are appropriate, and that the decision making is informed.

2.3.3 Ensure Endorsement of the Risk Management Process

Communication and consultation support decision making, and aim to give decision makers and other stakeholders a sense of responsibility about the management of risks. Risk communication should furthermore help ensure mutual understanding among decision makers and stakeholders, thereby avoiding that bad decisions are made due to misunderstanding and lack of information. More fundamentally, good procedures for communication and consultation help to ensure endorsement of and support for the risk management process as such.

Effective and efficient management of risk requires decision makers, stakeholders, and any key personnel to pull in the same direction. For this purpose it is important to achieve a common agreement on and mutual understanding of how risk should be managed.

2.3.4 Communicate Risk Assessment Results

The results of the risk assessment are an important part of the information that must be communicated to all relevant stakeholders. This will support decision making, improve the understanding of the sources and nature of risk, strengthen risk awareness, and generally make the organization better positioned for managing risk. The communication of the risk assessment results will help both internal and external stakeholders to understand decisions and prioritizations regarding the management of risk. The risk assessment results may also be important for demonstrating, for example, policy adherence or compliance with directives and regulations. The risk assessment results are also important for justifying treatment plans, including the required resources for risk mitigation. Communicating the results helps those with a vested interest to understand the basis on which decisions are made and why particular actions are required. This, in turn, helps to ensure endorsement of risk treatment plans from key stakeholders.

2.4 Risk Assessment

By risk assessment we mean activities aiming to understand and document the risk picture for specific parts or aspects of a system or an organization. The assessment includes the estimation of the risk level, as well as the identification of options for risk treatment. The results serve as a decision basis for risk management, including the decision of which controls and measures to implement to mitigate risk. The *risk assessment process*[1] is divided into five steps, as illustrated by Fig. 2.4.

2.4.1 Context Establishment

The context establishment is the preparatory step for the subsequent activities and involves the documentation of both the external and the internal context of relevance for the assessment in question. This step defines the goals and objectives of the risk assessment, and therefore requires the participation of decision makers. The *external context* includes the relationships with external stakeholders, as well as the relevant

---

[1] This is a slight deviation from ISO 31000. See Sect. 2.6 for an explanation.

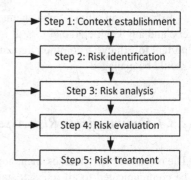

Fig. 2.4 Risk assessment
process

societal, legal, regulatory, and financial environment. The *internal context* includes
the relevant goals, objectives, policies, and capabilities that may determine how risk
should be assessed.

In addition to establishing this general context for the risk assessment, the *context
establishment* involves providing all the input that is needed for the following steps
of risk assessment. We refer to this as the *context description*, the contents of which
are discussed in the following. The *goals and objectives* are what we seek to achieve
by the risk assessment. These can be of a high level, such as the achievement of
business objectives or the provisioning of business services, but are important in
order to understand the target, scope, and focus of the assessment.

Definition 2.9 The *target of assessment* is the parts and aspects of the system
that are the subject of the risk assessment.

Definition 2.10 A *system* is a set of related entities that forms an integrated whole
and has a boundary to its surroundings.

Notice that our definition of system is very broad. An organization, for instance,
may be understood as a system according to this definition. The target of assessment (or target for short) includes the activities, processes, personnel, users, and all
other relevant entities constituting the subject of the risk assessment. During the risk
assessment we do the risk identification based on the description of the target. It is
therefore important that the description is at a level of abstraction that matches the
level of detail at which we aim to do the risk identification. It is useful also to decide
and explicitly specify the desired scope and focus of the assessment. The *scope of
the assessment* is the extent or range of a risk assessment; it defines what is held
inside of and what is held outside of the assessment. The *focus of the assessment* is
the main issue or central area of attention in a risk assessment; the focus is within
the scope of the assessment.

Together with the description of the target of assessment we need to specify our
assumptions about the target and its environment. An assumption is something we
take for granted or accept as true about the system in question, and the risk assessment is valid only given these assumptions. Examples of assumptions could be that

risks are caused by internal personnel only because we have been asked to restrict our attention to company personnel, or that certain service level agreements (SLAs) will be fulfilled by external suppliers. Assumptions are made to focus the risk assessment and avoid duplicating work. The reason for the above SLA assumption could be that the party in question conducts (or has recently conducted) in parallel a separate risk assessment addressing the potential impacts of unfulfilled SLAs. The documentation of all such assumptions is essential because they are needed as input to the risk assessment, and because the results of the assessment are valid only under these assumptions.

A crucial step in the context establishment, and in defining the focus of the assessment, is the identification and documentation of the assets with respect to which the risk assessment is conducted. Before we can do the asset identification, we need to be specific about who the party of the risk assessment is. What is held as assets, how critical, important, or valuable the assets are, and the degree to which they require protection can be determined only by considering the party. A risk assessment is typically conducted with respect to one party, but it is possible to allow for two or more.

Having specified the target and assets we can define the risk scales and the risk evaluation criteria. For defining the risk scales we need scales for consequences and likelihoods. In principle we can use the same consequence values for all kinds of assets, for example in terms of monetary loss. However, this can be challenging in a practical setting where it may be very hard to know the economic implications of risks. Therefore it is often more useful to describe consequences that are specific to the asset in question. For example, for availability of a service the consequences could be given in terms of downtime. For each asset we therefore first consider the nature and kind of consequences that can occur and how they will be measured. Moreover, because the same risk assessment may involve assets of different kinds, we may need to define several consequence scales, one for each kind of asset.

For the documentation of likelihoods we need only one scale, but we need to decide how the likelihoods shall be measured. Sometimes it is suitable to use general terms such as "seldom" or "often," and other times we use numeric, discrete scales. For some risk assessments the most suitable alternatives are frequencies or probabilities. The consequence and likelihood scales we define can be quantitative or qualitative, and they can be continuous, discrete, or given as intervals. In Chap. 12 we discuss the different alternatives more closely and give advice on which kind of scale to use for which purposes.

Risk levels are given by a function from likelihoods and consequences. This can be a mathematical function, for example by the multiplication of probability and monetary loss. In such a case of quantitative and continuous consequence and likelihood scales, the scale for risk levels is also quantitative and continuous. A more common way of specifying the risk function is by using a risk matrix with the likelihoods on one axis and the consequences on the other. Each cell then corresponds to a specified risk level. What is important when defining the risk function is that it serves as a basis for defining the risk evaluation criteria and that risk assessors and decision makers can distinguish between risk levels when the difference is significant for the risk evaluation. The risk matrix usually serves this purpose since we can always adjust the granularity by increasing or decreasing the number of cells. In Chap. 11 we present alternative ways of expressing risk and risk levels.

The *risk evaluation criteria* are the terms of reference by which the significance of risk is assessed. Because assets may be of different kinds and significance, we may need to define different evaluation criteria for different assets or different kinds of assets. The context description, which constitutes the collection of the above, serves as the input to and the basis for the risk assessment.

2.4.2 Risk Identification

By *risk identification* we mean activities aiming to identify, describe, and document risks and possible causes of risk. To this end we keep in mind two things. First, according to Def. 2.1, a risk is always associated with an incident. Second, there are three elements without which there can be no risk, namely asset, vulnerability, and threat. Without assets there is nothing to harm, without vulnerabilities there is no way to cause harm, and without threats there are no causes of harm. We therefore conduct the risk identification with respect to the identified assets by identifying threats and understanding how the threats may lead to incidents (and thereby risks) by exploiting vulnerabilities.

Definition 2.11 A *vulnerability* is a weakness, flaw, or deficiency that can be exploited by a threat to cause harm to an asset.

Examples of vulnerabilities are weak window lock and lack of intruder alarm, both of which a burglar can exploit in a break-in. Other examples are broken smoke detectors, insufficient staff training, and lack of back-up copies of critical operator manuals. The criticality of a vulnerability depends on the threats that may exploit them.

Definition 2.12 A *threat* is an action or event that is caused by a threat source and that may lead to an incident.

Threats may lead to incidents, but in order to identify threats and understand how they arise, we need to understand their initial causes, namely the threat sources.

Definition 2.13 A *threat source* is the potential cause of an incident.

A threat source can be human or non-human, and it can be tangible or intangible. Examples of human threat sources are burglars and negligent employees, while natural causes such as lightning or flood are non-human threat sources. Malware is an example of an intangible threat source.

Figure 2.5 illustrates how a threat source causes a threat that can lead to a risk by exploiting vulnerabilities. The arrow pointing backwards illustrates that threats can lead to other threats that eventually cause risks. During the risk identification we seek to understand and document how this can happen with respect to the identified assets. In practice we often structure the risk identification by starting at one end

Threat source   Threat   Vulnerability   Risk

Fig. 2.5 Threat sources cause risks

and working our way to the other end, for example, by first identifying potential incidents and then trying to understand how and why they can arise. As illustrated by Fig. 2.6, we can go back and forth while gradually building the risk picture. For example, a threat that we identify for a given incident can trigger the identification of other incidents. As explained further in Sect. 5.3, where to start and in which order to address the questions depend on the kind of risk we are dealing with. When con-

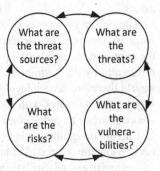

What are the threat sources?

What are the threats?

What are the risks?

What are the vulnera-bilities?

Fig. 2.6 Risk identification

ducting risk identification in a systematical manner, we need techniques for doing the identification, and we need suitable formats for describing and documenting the results. Techniques for risk identification include brainstorming, interviews, checklists, statistics, and approaches for gathering historical data. Risk assessors may also use modeling techniques such as event trees [24], Bayesian networks [5], attack trees [70], CORAS diagrams [47], or threat modeling [75] to support the description of risks and how they are related to threats and threat sources.

Which technique to use depends on a variety of factors, such as the desired level of detail, the available resources and the expertise and experience of the risk assessors. The same is the case for the choice of documentation format. Apart from plain prose, there are basically two formats for the description and documentation of risks, namely tables and graphical models. Graphical models, such as those mentioned above, are often designed for specific purposes, for example to support brainstorming, to explore causes and/or consequences of incidents and scenarios, or to facilitate more rigorous assessment. Tables are suitable for structuring the information in a systematic way, but are typically used for more high-level risk assessments.

Whatever techniques and level of detail we choose for the risk identification and documentation, we always need to make sure that we describe all the elements of the risk picture that we need for the purpose and objectives of the risk assessment. At the very least the documentation should include threat sources, vulnerabilities, risks, and assets.

2.4.3 Risk Analysis

By *risk analysis* we mean activities aiming to estimate and determine the level of the identified risks. As defined in Sect. 2.1, the risk level is derived from the combination of the likelihood and consequence. The objective of this step, therefore, is to estimate likelihoods and consequences for the identified incidents using the scales defined during the context establishment. An incident represents one risk for each of the assets it harms, and we need to estimate the consequence for each of these assets.

The impact or severity of an incident can be determined only by considering the party in question. The severity of a wrongly addressed postal letter that leads to the exposure of confidential patient information, for example, is likely to be judged differently by the hospital and the patient in question. The consequence estimation should therefore be conducted by a walk-through of all identified incidents and assigning the estimates with the involvement of personnel representing the party or someone who can judge consequences on behalf of the party.

Likelihood estimation is to determine the frequency or probability of incidents to occur using the defined likelihood scale. This requires the use of techniques for gathering empirical data. Such techniques include interviews and brainstorming sessions to gather expert opinions, inspection of logs or other statistical and historical data, and the use of available repositories. Many of the risk-modeling techniques such as Bayesian networks, attack trees, and CORAS diagrams, also come with support for likelihood estimation and documentation. How we choose to model or document the risks during the risk identification may therefore have some implications on which techniques are available for the likelihood estimation.

The desired level of detail of the risk assessment and documentation is another factor. Sometimes we are only interested in the likelihoods of the incidents and make our best estimates directly for these events. Very often, however, we need to understand how risks are most likely to arise, and which threat sources are most important. In that case we should also try to estimate the likelihood that threat sources initiate threats, and the likelihood that such threats may lead to incidents. This information will not only help to understand the most important threat sources and vulnerabilities, but also to determine the likelihood of the resulting incidents.

Once we have estimated the likelihood and consequences for each incident, we calculate the risk level of all identified risks by using the risk function we defined during the context establishment.

2.4.4 Risk Evaluation

By *risk evaluation* we mean activities involving the comparison of the risk analysis results with the risk evaluation criteria to determine which risks should be considered for treatment.

In principle this step is quite straightforward given the risk estimates and evaluation criteria. For example, if we have specified the risk evaluation criteria using the risk matrix, we simply need to plot each risk into the matrix to determine the risk level. However, because the risk evaluation is a decision point in the overall risk assessment process, we take the time to confirm the risk evaluation criteria and consolidate the risk estimates. Decision makers and other personnel that are involved in the risk assessment often gain new insight and knowledge about the risks and their consequences, and we must therefore make sure that the initially defined criteria are still appropriate. For the consolidation of the risk assessment results we focus on the risk estimates that we are uncertain about, and where this uncertainty implies doubt about the actual risk level.

We moreover need to investigate the identified risks to see whether certain sets of risks should be aggregated and evaluated as a single risk. This is to avoid the pitfall of accepting a set of risks that individually are non-critical, yet unacceptable in combination. Even if the likelihood and consequence of the respective risks yield an acceptable risk level for both, it may be that the two are unacceptable taken together. Another situation is when we have a number of separate incidents that harm the same asset, and where the incidents can be understood as special cases of the same, more general, incident, or where the incidents may be caused by the same threat. How to aggregate likelihoods and consequences depends on the kinds of scales we use and any statistical dependencies between incidents.

A final recommendation for the risk evaluation is to group risks that have elements in common. Risks that share threat sources, threats, vulnerabilities, and/or assets may often be treated by the same means. Therefore, in preparation for the risk treatment and to facilitate cost-efficient treatment, we go through the identified risks and group them as we see appropriate.

2.4.5 Risk Treatment

By *risk treatment* we mean activities aiming to identify and select means for risk mitigation and reduction. Sometimes this step is referred to as risk modification to stress the fact that risks can both decrease and increase as a consequence of treatments. In particular, this is the case for approaches to risk management that may involve taking or increasing risk in order to pursue an opportunity. In this book we focus only on the identification of treatments for the purpose of reducing or removing risks. This is reflected by the following definition.

Definition 2.14 A *treatment* is an appropriate measure to reduce risk level.

In principle we should seek to treat all risks that are unacceptable, but in the end this is a question of cost and benefit, no matter the risk level. If a low risk is very cheap to eliminate, we might do so even if the risk in principle is acceptable. And, similarly, if the cost of treating a very high risk is unbearable there may be no other option than to accept it.

The risk treatment activity, therefore, should involve both the identification and the analysis of treatments. The treatment identification can be done similarly to the risk identification, for example via brainstorming or by the use of available lists and repositories. The selection of which treatments to implement should be the result of an analysis of the costs and benefits of the identified treatments. The analysis should take into account that some treatments can create new risks, and that some groups of treatments can reduce the isolated effect of each other. For example, unauthorized access may be mitigated by improved intrusion detection or by stronger access control, but we cannot expect the effect of the two in combination to be the sum of the effects of each of them alone.

There are four main options for risk treatment, namely risk reduction, risk retention, risk avoidance, and risk sharing [32]. We may reduce risk by reducing the likelihood and/or consequence of incidents. To do this we seek options to remove threat sources, remove or reduce the severity of vulnerabilities, or reduce the likelihood of threats by other means. Risk retention is to accept the risk by informed decision. This is typically an option for risks that are acceptable according to the risk criteria, or risks that are too costly to treat given the alternative options. Risk avoidance is simply to avoid the activity that gives rise to the risk in question, which sometimes is the only option for unacceptable risks. Risk sharing is to transfer the risk or parts of it to another party, for example, by insurance or sub-contracting.

2.5 Monitoring and Review

Monitoring and review apply both to the underlying risk management framework and to the risk management process, but specifically also to the identified risks and to the measures that the organization implements in order to treat risks. *Monitoring* is the continual checking, supervising, critically observing, or determining the current status in order to identify deviations from the expected or required status. The *review* activity is to determine the suitability, adequacy, and effectiveness of the risk management process and framework, as well as risks and treatments. The main purposes of the monitoring and review process are as follows [25]:

- Ensure that controls are effective and efficient
- Obtain further information to improve risk assessment
- Analyze and learn lessons from incidents, changes, trends, successes, and failures
- Detect changes
- Identify emerging risks

2.5.1 *Monitoring and Review of Risks*

Risks are not static and must therefore be monitored and reviewed. This includes all aspects of risks, including assets, threats, and vulnerabilities, as well as likelihoods and consequences. Constant monitoring is necessary for detecting and identifying changes to any of these aspects. Existing risk assessment results and other risk documentation must be reviewed to determine whether they are still valid. The monitoring and review of risk serves as a basis for taking actions, such as modifying the risk picture or conducting new risk assessments. Elements to monitor include the following:

- Assets: The set of assets that are of concern in the overall management of risk must be monitored in order to determine whether there are significant changes in asset value or priority over time. Changes in the internal or external context may moreover introduce new assets and make others obsolete.
- Threats: Internal or external changes could introduce new threats, including changes of assets or asset values. In some cases specific and known threats can be observed directly. In other cases it may be required to conduct risk assessments in order to thoroughly identify new threats.
- Vulnerabilities: Known vulnerabilities can be monitored in order to determine those that potentially could be exposed to new threats. They can also be monitored to detect changes, such as vulnerabilities that become more easy to exploit or more widespread.

Previous risk assessments are an important source of risk factors that should be monitored. In particular, this is the case for residual and acceptable risks that over time could evolve.

2.5.2 *Monitoring and Review of Risk Management*

Organizations also need to conduct continuous monitoring and review of the risk management framework and process. This is to ensure that the framework and process, as well as all related activities, procedures, roles, and responsibilities, remain relevant, appropriate, and adequate for the organization. Moreover, the review activity is conducted to verify that the risk evaluation criteria are valid over time, and that they are consistent with policies and business objectives.

Generally, any changes in the internal or external context that may affect the adequacy of the risk management need to be monitored and reviewed. This may include the following:

- Legal and environmental context
- Competition context
- Assets and asset values
- Risk evaluation criteria
- Resources required for adhering to the risk management framework

The risk management monitoring and review may result in changes in assets or evaluation criteria. But the required changes may also be more profound, for example, by changing the risk assessment techniques or tools, or by changing risk management procedures and responsibilities.

2.6 Further Reading

The terminology introduced in this chapter is largely based on the risk management vocabulary of ISO Guide 73 [26]. The presentation of the risk management process and how this process relates to the risk management framework and principles is based on the ISO 31000 risk management standard [25]. This standard also makes use of the vocabulary of ISO Guide 73.

Note that ISO 31000 refers to risk assessment as the three activities of risk identification, risk analysis, and risk evaluation. In this book we use the term "risk assessment" in a broader sense. It also include the activities of context establishment and risk treatment. There are two reasons for this: First, ISO 31000 offers no term denoting the process consisting of these five activities. Second, in our view, this better reflects how the term "risk assessment" is used in practice.

In addition to these ISO standards, we refer the interested reader to the ISO/IEC 27005 [32] standard on information security risk management. This standard is much more limited than ISO 31000 as it concerns information security risks, but because it builds closely on the latter, it gives some good insights into many principles of risk management in general.

For a useful and quite comprehensive overview and classification of risk assessment techniques, the reader is referred to IEC 31010 [30]. The overview includes techniques for risk identification, risk analysis, and risk evaluation.

Chapter 3
Cyber-systems

How organizations should conduct risk management largely depends on the kind and nature of the systems of concern. In this book we are concerned with systems that make use of a cyberspace, namely cyber-systems, as further elaborated below.

3.1 What is a Cyberspace?

The most prominent example of a cyberspace is the Internet, but we must be careful not to confuse the two terms as they are not interchangeable. We define cyberspace as follows.

> **Definition 3.1** A *cyberspace* is a collection of interconnected computerized networks, including services, computer systems, embedded processors, and controllers, as well as information in storage or transit.

For most organizations and other stakeholders, cyberspace is for all practical purposes synonymous with the Internet, which is a global cyberspace in the public domain [28, 78]. Our definition, however, is more general: Any collection of interconnected networks [78] is a cyberspace. A common form of interconnected computerized networks is a collection of local area networks (LANs) that are connected by a wide area network (WAN). Examples of cyberspaces that are not connected to the Internet are military computer networks, as well as emergency communication networks and systems. Examples of cyberspaces that preceded the Internet were the non-commercial National Science Foundation Network (NSFNET), as well as the Advanced Research Projects Agency Network (ARPANET) that was operative from 1969.

© The Author(s) 2015
A. Refsdal et al., *Cyber-Risk Management*, SpringerBriefs
in Computer Science, DOI 10.1007/978-3-319-23570-7_3

3.2 What is a Cyber-system?

In order to understand risk in relation to a cyberspace, we need to understand and take into account the scope of the subject matter. Risks that somehow stem from or are due to a cyberspace, such as the Internet, may obviously have implications well beyond the cyberspace alone. This is because any system that in one way or another depends on a cyberspace may also be vulnerable due to this dependency. To account for this scope, we introduce the notion of cyber-system.

Definition 3.2 A *cyber-system* is a system that makes use of a cyberspace.

A cyber-system may include information infrastructures, as well as people and other entities that are involved in the business processes and other behavior of the system. This means that cyber-systems are part of the organizational structure of most organizations.

Cyber-systems have moreover become more and more ubiquitous in society at large. Citizens, enterprises, governments, and a range of other stakeholders rely on software systems and Internet connectivity for the provisioning and consumption of services. Such services include welfare, health, banking, entertainment, social networks, trade, energy, transportation, and so on. Many of the systems that are critical for society at large, so-called critical infrastructures, are also cyber-systems. Such infrastructures include, for example, telecommunication, transportation, finance, power supply, water supply, and emergency services.

A cyber-physical system is a special case of a cyber-system that interacts with its physical surroundings.

Definition 3.3 A *cyber-physical system* is a cyber-system that controls and responds to physical entities through actuators and sensors.

Cyber-physical systems are increasingly part of our daily lives, and their networked sensors and actuators are used to control smart grids, smart homes, production lines, automotive controllers, and other kinds of entities.

The UML class diagram of Fig. 3.1 shows how the notions we have introduced in this chapter relate to each other. The white-headed arrow pointing from cyber-system to system means that cyber-system should be understood as an instance of the more general notion of system. The diagram moreover expresses that a cyber-physical system is a cyber-system, and therefore also a system. Cyber-systems, including cyber-physical systems, are always related to a cyberspace.

3.3 Further Reading

There is a plethora of textbooks and other literature on the topics of cyberspace and cyber-systems, and sometimes the terms are given various meanings.

For a thorough introduction to computer networks in general see, for example, the textbook by Andrew Stuart Tanenbaum [78]. Both this textbook and others, such

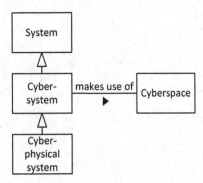

Fig. 3.1 Cyber-system concepts

as the ISO/IEC 27032 standard [28], provide definitions of the synonyms internetwork and internet. As in this book the capitalized term Internet refers to the global, worldwide Internet, one of several internetworks.

ISO/IEC 27032 defines cyberspace as an environment within the Internet. The standard moreover limits its scope to the cyberspace alone, not considering systems that make use of the cyberspace. It is also limited to the purely virtual and non-tangible aspects.

The notion of cyberspace is used in many other contexts. The EU uses it more or less in the same meaning as the Internet [13]. The International Telecommunication Union (ITU) [37] uses the term cyber environment, which roughly equals to our notion of cyber-system. Other initiatives focus more on cyberspace in relation to critical infrastructures [55, 80].

Cyber-physical systems receive much attention within industry as well as research [86, 67, 19] and require careful design considerations [44].

Chapter 4
Cybersecurity

In this chapter we define and explain the notion of cybersecurity. What characterizes cybersecurity, and what are the kinds of threats that cybersecurity must prevent or provide protection from? We also explain how cybersecurity relates to information security, critical infrastructure protection, and safety.

4.1 What is Cybersecurity?

While cybersecurity may involve the security of a cyberspace itself, most organizations are concerned with the protection of their own cyber-systems from cyber-threats. Both of these concerns are within the scope of our definition of cybersecurity.

> **Definition 4.1** *Cybersecurity* is the protection of cyber-systems against cyber-threats.

Cyber-threats are those that arise via a cyberspace, and are therefore a kind of threat that any cyber-system is exposed to.

> **Definition 4.2** A *cyber-threat* is a threat that exploits a cyberspace.

Cyber-threats may be malicious or they may be non-malicious. Examples of malicious threats are denial of service (DoS) attacks and injection attacks that are caused by intention. Non-malicious threats are, for example, systems that crash due to programming errors or loss of Internet connection due to wear and tear of communication cables or other hardware.

Notice, importantly, that what defines cybersecurity is not what we seek to protect, but rather what we seek to protect it from; it is not defined by the kinds of assets that are to be protected, but rather by the kinds of *threats* to assets.

© The Author(s) 2015
A. Refsdal et al., *Cyber-Risk Management*, SpringerBriefs
in Computer Science, DOI 10.1007/978-3-319-23570-7_4

The assets of concern depend on the organization and the cyber-system in question, although cybersecurity is often about the protection of information assets or infrastructure assets. We discuss this more closely in the following sections by relating cybersecurity to information security, infrastructure protection, and safety.

4.2 How Does Cybersecurity Relate to Information Security?

Information security is the preservation of confidentiality, integrity, and availability of information [35]. Information can come in any form, be it electronic or material, or even as the knowledge of personnel. In order to ensure and maintain information security, information in all formats needs to be protected from threats and threat sources of any kind, including physical, human, and technology-related threats. Cybersecurity, on the other hand, concerns protection from threats that use a cyberspace. Such threats may target information assets, which is why information security is an important part of cybersecurity. However, cybersecurity addresses only those information assets that can be targeted via a cyberspace. Cybersecurity is not limited to the protection of information assets alone. As we discuss below, it often concerns the protection of infrastructure. We may also be concerned about the wider impact of threats to information or infrastructure security in order to protect assets such as life, health, reputation, revenue, and so forth.

Most standards and guidelines on cybersecurity relate cybersecurity to information security. This is to be expected as there is considerable overlap, but in order to properly understand what cybersecurity is and how to ensure it, we must be careful not to confuse these two kinds of security. Cybersecurity goes beyond information security in that it is not limited to the protection of information assets and the preservation of confidentiality, integrity, and availability of information. Information security, on the other hand, goes beyond cybersecurity in that it is not limited only to threats that arise via a cyberspace.

4.3 How Does Cybersecurity Relate to Critical Infrastructure Protection?

Infrastructure security, in particular *critical infrastructure protection* (CIP) and critical information infrastructure protection (CIIP), is concerned with the prevention of the disruption, disabling, destruction, or malicious control of infrastructure [12, 28]. Such infrastructures include, for example, telecommunication, transportation, finance, power supply, water supply, and emergency services. CIP is crucial for societal security, as well as for organizations and other stakeholders that provide or rely on critical infrastructures.

Many critical infrastructures make use of a cyberspace and are therefore cyber-systems. Hence, the security of such systems involves protection from cyber-threats.

CIP in general, however, goes beyond cybersecurity since CIP involves the protection and the security of any critical infrastructures, whether or not they make use of a cyberspace. Cybersecurity, on the other hand, concerns the protection of infrastructures that can be targeted via a cyberspace. Such infrastructures include, for example, telecommunication networks and cyber-physical systems like a smart grid.

How cybersecurity relates to information security and CIP is illustrated in the Venn diagram of Fig. 4.1. From the diagram we see that while cybersecurity may involve both information security and CIP, the former is not simply a combination of the latter two.

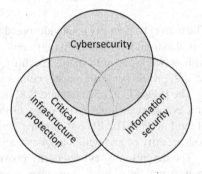

Fig. 4.1 Cybersecurity vs. information security and critical infrastructure protection

4.4 How Does Cybersecurity Relate to Safety?

Safety can be defined as the protection of life and health by the prevention of physical injury caused by damage to property or to the environment [1, 23]. One of the main differences between safety and cybersecurity is that while safety focuses on system incidents that can harm the surroundings, cybersecurity focuses on threats that cause harm via a cyberspace. A further difference is that the assets that are considered with respect to safety are usually limited to human life and health, as well as environmental assets, while the assets of concern with respect to cybersecurity can be anything that needs to be protected.

The distinction between safety and cybersecurity does not mean that safety issues are outside the scope of the latter. The reason for this is that safety incidents may have security impact, in the same way that security incidents may have safety impact. For example, a cyber-attack on a power distribution control system that leads to a blackout could have fatal safety consequences for hospital patients. And a safety incident, such as a gas explosion, could damage information systems and disable security controls, thereby leaving a system vulnerable to cyber-threats. When seeking to ensure cybersecurity we therefore need to take into account safety incidents that may yield vulnerabilities or that otherwise can be exploited by threat sources.

How cybersecurity relates to safety is illustrated by the Venn diagram of Fig. 4.2.

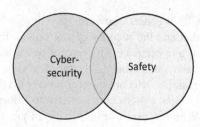

Fig. 4.2 Cybersecurity vs. safety

4.5 Further Reading

The term cybersecurity is in widespread use. As it is used in many different contexts, it is also used with somewhat different meanings. Some use it quite synonymously with network and information security, others focus more strictly on information security, while some are mostly concerned with CIP.

The ISO/IEC 27032 standard [28] defines cybersecurity as information security in a cyberspace, limiting its scope to the strictly virtual and non-physical aspects of the Internet. The EU has ongoing activities regarding cybersecurity that concern both security of and within a cyberspace [13], as well as CIIP [12].

The definition of cybersecurity provided by the ITU includes both information security and the protection of cyber-systems [37]. Others focus more strictly on CIP, such as the National Institute of Standards and Technology (NIST) cybersecurity framework [55].

The definition of cybersecurity provided by the Committee on National Security Systems (CNSS) [7] is similar to our definition, although the former is restricted to adversaries and attacks, rather than cyber-threats in general.

There are a lot of standards and literature on safety. As a detailed discussion of this term is outside the scope of this book, we refer the reader to the IEC/TR 61508-1 standard [23] for a common definition. Algirdas Avižienis et al. [1] discuss how safety relates to concepts such as security and dependability.

Chapter 5
Cyber-risk Management

In this chapter we specialize risk management to the domain of cyber-systems. We highlight what is special about cyber-systems and cyber-threats from a risk management perspective, focusing in particular on the nature of cyber-risks and the options and means we have for managing them. First we explain what we mean by cyber-risk. Thereafter we specialize the three main processes of risk management to cope with cyber-risk.

5.1 What is Cyber-risk?

Cyberspace has considerable impact on the kind and nature of the threats and the risks that may appear, as well as on the procedures and techniques to conduct risk management and risk assessment. One striking aspect of cyberspace is that it is potentially extremely far-reaching. This means that the possible threat sources can reside anywhere in the world, yet with the potential of causing damage deep inside the cyber-system of our concern. Another crucial aspect is that a substantial share of cyber-threats are malicious; they are caused by adversaries with motives and intentions. On the other hand, there are also non-malicious cyber-threats.

Cyber-risk management is concerned with risks caused by cyber-threats, which motivates the following definition.

Definition 5.1 A *cyber-risk* is a risk that is caused by a cyber-threat.

Although we are concerned with cyber-systems, it is important to understand that cyber-risk is not the same as any risk that a cyber-system can be exposed to; cyber-risks are limited to the risks that are caused by cyber-threats. The risk of a server on which our cyber-system is running being damaged by water flooding, for example, is not a cyber-risk unless a cyber-threat is a contributing factor. Confidentiality breaches due to virus attacks via cyberspace and loss of availability due to DoS attacks, however, are examples of cyber-risks.

© The Author(s) 2015
A. Refsdal et al., *Cyber-Risk Management*, SpringerBriefs
in Computer Science, DOI 10.1007/978-3-319-23570-7_5

Next, in order to understand the nature of cyber-risks and how to manage them we distinguish between *malicious cyber-risk* and *non-malicious cyber-risk*. We say that a cyber-risk is malicious if it is (at least partly) caused by a malicious threat, and non-malicious otherwise.

Notice, importantly, that by this definition some cyber-risks are both malicious and non-malicious. These are cyber-risks that can be caused by either a malicious threat or a non-malicious threat. Consider, for example, an incident of unauthorized access to some sensitive data. A potential occurrence of this incident as caused by a hacker is a malicious cyber-risk, while a potential occurrence that is caused by accidental posting of the data on an open website is a non-malicious cyber-risk.

There are also incidents that happen only due to the combined occurrence of a malicious and a non-malicious threat. An example of this is an intrusion that occurs while the intrusion detection and prevention system is down due to an accidental failure. We classify these as malicious cyber-risks since they cannot occur without the malicious threat.

The Venn diagram of Fig. 5.1 provides a summary: Cyber-risks are the union of malicious and non-malicious cyber-risks, and cyber-risks are only a subset of the risks that cyber-systems can be exposed to. Moreover, the intersection between malicious and non-malicious cyber-risk represents the cyber-risks that can be caused by either a malicious threat or a non-malicious threat.

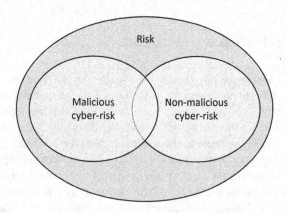

Fig. 5.1 Malicious and non-malicious cyber-risk

5.2 Communication and Consultation of Cyber-risk

The process of communication and consultation described in Sect. 2.3 for risk management in general is equally suited to the more narrow domain of cyber-risk. There are, however, certain issues imposed by cyberspace that require particular attention. First, due to the nature of cyberspace, cyber-systems may potentially have stakeholders everywhere. These stakeholders may be consumers of services or information provided by the cyber-system of our concern, or they may be providers of services to this cyber-system. It is important to consider all stakeholders, both individuals and organizations, when determining relevant sources of information and identifying who may be affected by cyber-risks. We moreover need plans and procedures for how to provide, share, obtain, and make use of the information of relevance.

Second, also due to cyberspace, there may potentially be adversaries everywhere, and any major incident somewhere in the world may have considerable impact on our cyber-system. Coping with these numerous parameters requires increased focus on information collection by monitoring and surveillance.

For the process to be efficient it is necessary to establish a classification and categorization of information. For the purpose of representing and understanding relevant information, organizations may use established standards or repositories, see Sect. 5.5, or they may define their own. The objective is to maintain a repository of up-to-date information regarding, for example, cyber-threats, vulnerabilities and incidents, potential and confirmed adversary profiles, current strategies and mechanisms for cyber-risk mitigation, and so forth. The classification and categorization may also include characterizations of cyber-systems, including, for example, assets and cyber-system profiling.

For many organizations it is essential to establish communication procedures for handling major incidents. Efficient communication, for example via a public relations team, is often an important element of good incident response planning.

5.3 Cyber-risk Assessment

There are two things in particular that distinguish risk assessment in the context of cyber-systems from the general case. First, the potentially far-reaching extent of a cyberspace implies that also the origins of threats are widespread, possibly global. Second, the number of potential threat sources and threats, both malicious and non-malicious, is very large. In combination this means that the search area and the number of sources of potentially relevant information about cyber-risk are extremely large and may seem overwhelming. We therefore need procedures and techniques that provide guidance and direction.

Figure 5.2 shows the specialization of the risk assessment process to cyber-systems. The most obvious difference from the general case is that the risk identification step is divided into two separate steps: Step 2a focusing on malicious cyber-risks and Step 2b focusing on non-malicious cyber-risks. We make this distinction because the nature of threats, threat sources, and vulnerabilities, and how to approach their identification, is highly dependent on whether we are dealing with malicious intent or not.

Human adversaries who deliberately and actively seek to cause harm are hard to predict, and the consequences of the incidents they cause can be difficult to estimate. We basically assess a game. There are two opponents with opposing goals: The

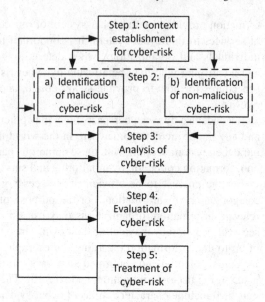

Fig. 5.2 Process for cyber-risk assessment

adversary (the malicious threat source) who actively seeks to harm assets, and the defender (the system owner or the party on whose behalf we do the assessment) who tries to prevent this from happening. The aim of Step 2a is to identify risks based on the potential ways in which such a game can play out. The motives, intentions, abilities, skills, resources, and so forth of the adversary are essential in this context. A good starting point in the identification of cyber-risks caused by malicious threats is therefore the identification and characterization of potential threat sources.

Understanding how non-malicious threats arise, such as accidents and failures, on the other hand, is a different kind of challenge. There is normally little need to capture intent or motive for such threats. Moreover, because there are an almost unlimited number of ways unintentional things may happen, it easily becomes overwhelming if we start by identifying threat sources and threats. In conducting Step 2b we recommend instead to start from the assets and the ways in which they may be harmed. In this way we make sure that we focus strictly on what we seek to protect, and that we proceed in a manner that is both effective and efficient. In other words, by first asking what can go wrong and then asking how, we help ourselves to keep the right focus. If, instead, we started by asking how something could happen unintentionally or by accident and what could possibly be the cause, we would soon find ourselves moving in all kinds of directions.

In the rest of this section we describe the specialization of each step of the cyber-risk assessment process in more detail.

5.3.1 Context Establishment for Cyber-risk

What distinguishes the context establishment of a cyber-risk assessment from the general case is that we need to understand and document how the cyber-system in question makes use of and interacts with cyberspace. This gives a basis for understanding how and where cyber-threats arise, as well as which assets are relevant to focus on.

As part of the description of the target of the assessment, we therefore include the interface to and interaction with the cyberspace and other relevant parts of the environment. Understanding and documenting the interface to the cyberspace is important for cyber-risk management in general and for identification of cyber-risks in particular. The cyber-threats arise in or via the cyberspace, and the interface between the cyberspace and the target of assessment overlaps with the attack surface. The *attack surface* is all of the different points where an attacker or other threat source could get into the cyber-system, and where information or data can get out [60].

Typical assets of concern in the setting of cyber-risk assessment are information and information infrastructures, including software, services, and networks. However, in order to understand the wider implications of cyber-threats and incidents, we need to take into account assets that can be harmed as a further consequence. Relevant concerns in this respect are, for example, reputation, image, market share, revenue, and legal compliance. The latter is relevant regarding, for example, data protection and privacy. Moreover, although cyberspace and cyber-systems are typically associated with the virtual and the intangible, it is important not to limit the focus to such aspects alone. Cyber-threats and incidents can also cause physical harm, including harm to life, health, and the environment.

5.3.2 Identification of Malicious Cyber-risk

To identify malicious cyber-risk it is often helpful to think in terms of a game such as chess. As illustrated by the UML class diagram in Fig. 5.3, there are two players, namely an adversary (the opponent or malicious threat source) and a defender on whose behalf we are assessing. The defender is represented by the target and a set of assets. Our role as risk assessors is to observe and assess this game. In particular, we try to foresee what the future moves of the adversary might be and to provide advice on how to counter these moves. What we can expect from the adversary depends on the motives and the abilities of the adversary, as well as on the helpers and the resources available to the adversary.

The attack strategies of the adversary are typically conditional on the strengths and weaknesses of the defender. As illustrated by the lowermost ellipse in Fig. 5.4, we as assessors are supposed to deliver as output a risk model obtained by documenting and assessing how and to what extent the adversaries of relevance may exploit these weaknesses. As captured by the uppermost ellipse in Fig. 5.4, our input is the target description and the selected assets, both obtained from Step 1.

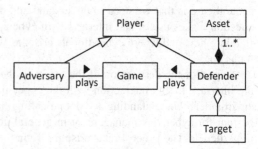

Fig. 5.3 Assessing the game between an adversary and a defender

The nature of the game obviously depends on who the defender is facing. As indicated by Fig. 5.4, we therefore start by identifying and documenting the properties of the potential adversaries, namely the malicious threat sources. When the threat sources have been identified and sufficiently documented, we proceed by investigating for each of them to what extent and in what way they may harm the assets. As illustrated by Fig. 5.4, we proceed via identification of malicious threats and the vulnerabilities these threats may exploit (called malicious vulnerabilities) to the identification of incidents. When we have completed the identification of malicious cyber-risk we document the results in a risk model which forms the output of the malicious risk identification. By *risk model* we mean any representation of risk information, such as threats, vulnerabilities, incidents, risks, and how they are related.

In practice, and as illustrated by the arrows in the figure, there may of course be exceptions from the ordering, as well as iterations back and forth, while identifying new elements. In the following we describe the contents of each step in further detail.

- *Malicious threat source identification:* To identify relevant and possible malicious threat sources we need to understand who may want to initiate attacks, what motivates them, what their capabilities and intentions are, how attacks can be launched, and so forth. We also need to take into account that although the malicious threat sources are often human, they may also be non-human such as a computer virus. There is motive and intent behind even non-human malicious threat sources because such threat sources are introduced deliberately and for a purpose. In principle we could choose to view the initial human actor as the threat source, but this depends on our target and scope and their relation to the threat in question. Typically, if malware has been developed specially to attack our target of assessment, then we view the developer of the malware as the initial threat source; otherwise, we view the malware itself as the threat source.

  Many malicious threat sources reside outside the cyber-system in question, but some are internal. To facilitate the identification it may be useful to consult rel-

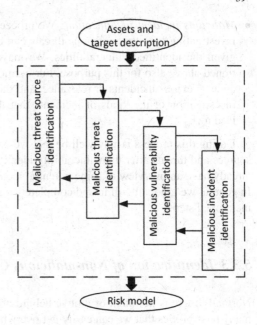

Fig. 5.4 Identification of malicious cyber-risk

evant sources such as international standards, annual and biannual reports on cybersecurity and cyber-threats, and open repositories.

- *Malicious threat identification:* For each of the malicious threat sources we proceed by identifying the malicious threats it may initiate that in some way or another may harm the identified assets. We pay particular attention to the interface to the cyberspace and the documented attack surface. In conducting this task we may involve people with first-hand knowledge about the target of assessment. Information can be gathered, for example, via questionnaires, interviews, workshops, and brainstorming sessions. We make active use of the description of the target of assessment, investigating where and how attacks can be launched. Examples of helpful catalogues and repositories that concern cybersecurity and cyber-threats in particular are those that are provided by MITRE [51, 52] and OWASP [61].
- *Malicious vulnerability identification:* In order to identify the vulnerabilities that the malicious threats may exploit we still focus on the identified attack surface. Additionally, we investigate existing controls and defense mechanisms to determine their strength and adequacy with respect to the identified threats and assets. As before we may consult system users and other personnel, as well as open information sources. For specific threats or vulnerabilities we may also conduct various kinds of security testing, such as penetration testing and vulnerability scanning. Such testing can serve as a means to check whether or how easily a specific threat source can actually launch an attack. We can do testing also to investigate the severity of known vulnerabilities, search for potential vulnerabilities, and search for possible incidents that the malicious threats may lead to.

- *Malicious incident identification:* We proceed to the incident identification by
  investigating how the malicious threats can cause harm to the identified assets
  given the identified vulnerabilities. We may use most of the techniques mentioned above also for this purpose. Furthermore, event logs provide information
  about previous incidents of relevance, and the various means of testing help the
  investigation of the kinds of incidents that the threats and vulnerabilities may
  lead to.

During this process it may well be that we backtrack and identify further threat
sources and threats after the vulnerability identification, and it may also be that we
already have an overview of some potential incidents that we aim to assess further.
In general we gradually fill in and complete the risk model by revisiting the abovementioned steps.

5.3.3 Identification of Non-malicious Cyber-risk

Normally there is no intent or motive behind non-malicious risks and there are so
many possibilities that we can easily get overwhelmed. It is therefore normally not
practical to start by identifying and documenting threat sources. Instead, as illustrated by Fig. 5.5, we recommend starting from the valuables to be defended, namely
the identified assets, and then working outwards in the direction of the arrows. For
each asset, the initial question is in what way it may be directly harmed. Each possibility corresponds to an incident. Next, we proceed by identifying the vulnerabilities
and threats that may cause these incidents, focusing only on the parts and aspects
of the target that are of relevance to the identified incidents. Finally, we identify the
non-malicious threat sources that can cause the threats.

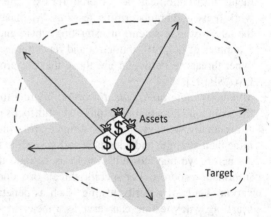

Fig. 5.5 Assessing how assets can be exposed to non-
malicious threats

This asset-driven process, as illustrated by Fig. 5.5, allows us to ignore all parts of
the target (the area with light shading) that are not relevant to the assets in question.

By starting from and strictly focusing on the assets, we make sure that we address only the relevant parts and aspects of the target (the area with darker shading). In other words, we use the assets to make the identification of non-malicious cyber-risks as efficient as possible.

The process is further illustrated by Fig. 5.6, where the initial step is the incident identification using the assets and the target description from the context establishment as input. For each incident, and as illustrated by the subsequent steps in the figure, we then proceed via vulnerabilities and threats to the identification of threat sources. We document the results during the process to produce the risk model that is the output of the risk identification, as illustrated by the ellipse at the bottom of the figure. Also here we may iterate back and forth, and deviate from the presented order when appropriate. In the following we describe the contents of each step in

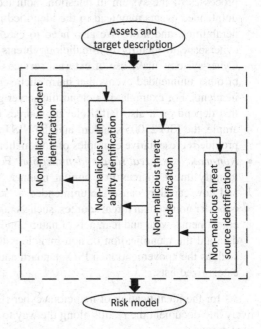

Fig. 5.6 Identification of non-malicious cyber-risk

turn, following the overall order as depicted in Fig. 5.6.

- *Non-malicious incident identification:* To identify incidents it is often useful to start by investigating how the assets are represented and how they are related to the target of assessment. For incidents with respect to information assets, for example, we investigate how the information is stored and processed in the system and in cyberspace, which applications and users have access to read or modify the information from where, how the information is transmitted, and so forth. For intangible assets, such as reputation, we need to understand how these are related to which parts or aspects of the target of assessment. Accidents and unintended acts are often recurring and known; we may therefore use logs, monitored data, and other historical data to support the identification.

- *Non-malicious vulnerability identification:* For the identification of vulnerabilities we may investigate technical parts of the target of assessment, as well as the culture, routines, awareness, and so forth of the organization and personnel in question. Relevant system properties that need to be investigated may, for example, be liberal access control, security mechanisms or barriers that are missing or that can be bypassed, and inconvenient application interfaces to the extent they open for accidental or unintended incidents to occur. Relevant issues regarding the organization and personnel include, for example, training, routines and procedures, and time pressure. Open sources, such as the ISO 27005 standard [32], come with lists of typical vulnerabilities.
- *Non-malicious threat identification:* In the identification of non-malicious threats we make use of the target description to systematically go through the uses and processes of the system in question, both technical and non-technical. Which unintended events may lead to the identified incidents due to the identified vulnerabilities, and how? We also need to carefully consider the interface to the cyberspace to identify non-malicious threats that arise outside of the system. Cyber-systems make use of external services and infrastructures, and accidents or other unintended events that harm such services or infrastructures can cause incidents. For example, if a cloud file server goes down, it may be that systems that depend on it also fail. Relevant sources on typical threats include, for example, the ISO 27005 standard and the NIST risk assessment guide [54], which provide representative examples of non-malicious threats.
- *Non-malicious threat source identification:* For each of the identified threats we identify threat sources in a similar manner. Who are the users of the system, and how can they cause the unintended or accidental events? We also need to consider non-human threat sources, such as failure of hardware or other technical components, wear and tear, acts of nature, and so forth. Event logs and historical data aid the identification of non-malicious threat sources, as do open sources such as the abovementioned ISO standard and NIST guide, which both provide categorized lists.

As for the identification of malicious cyber-risks, we conduct the process iteratively and document the results along the way to produce the risk model that is the final output.

5.3.4 Analysis of Cyber-risk

There are two aspects in particular that distinguish the analysis of cyber-risk from risk analysis in general. First, for malicious threats behind which there is human intent and motive, it can be hard to estimate the likelihood of occurrence. Second, due to the nature of cyber-systems we have several options for logging, monitoring, and testing that can facilitate the analysis. In addition to this there are various open resources that we can make use of.

Stochastic Modelling and Applied Probability
formerly: Applications of Mathematics

1 Fleming/Rishel, **Deterministic and Stochastic Optimal Control** (1975)
2 Marchuk, **Methods of Numerical Mathematics** (1975, 2nd. ed. 1982)
3 Balakrishnan, **Applied Functional Analysis** (1976, 2nd. ed. 1981)
4 Borovkov, **Stochastic Processes in Queueing Theory** (1976)
5 Liptser/Shiryaev, **Statistics of Random Processes I: General Theory** (1977, 2nd. ed. 2001)
6 Liptser/Shiryaev, **Statistics of Random Processes II: Applications** (1978, 2nd. ed. 2001)
7 Vorob'ev, **Game Theory: Lectures for Economists and Systems Scientists** (1977)
8 Shiryaev, **Optimal Stopping Rules** (1978)
9 Ibragimov/Rozanov, **Gaussian Random Processes** (1978)
10 Wonham, **Linear Multivariable Control: A Geometric Approach** (1979, 2nd. ed. 1985)
11 Hida, **Brownian Motion** (1980)
12 Hestenes, **Conjugate Direction Methods in Optimization** (1980)
13 Kallianpur, **Stochastic Filtering Theory** (1980)
14 Krylov, **Controlled Diffusion Processes** (1980)
15 Prabhu, **Stochastic Storage Processes: Queues, Insurance Risk, and Dams** (1980)
16 Ibragimov/Has'minskii, **Statistical Estimation: Asymptotic Theory** (1981)
17 Cesari, **Optimization: Theory and Applications** (1982)
18 Elliott, **Stochastic Calculus and Applications** (1982)
19 Marchuk/Shaidourov, **Difference Methods and Their Extrapolations** (1983)
20 Hijab, **Stabilization of Control Systems** (1986)
21 Protter, **Stochastic Integration and Differential Equations** (1990)
22 Benveniste/Métivier/Priouret, **Adaptive Algorithms and Stochastic Approximations** (1990)
23 Kloeden/Platen, **Numerical Solution of Stochastic Differential Equations** (1992, corr. 3rd printing 1999)
24 Kushner/Dupuis, **Numerical Methods for Stochastic Control Problems in Continuous Time** (1992)
25 Fleming/Soner, **Controlled Markov Processes and Viscosity Solutions** (1993)
26 Baccelli/Brémaud, **Elements of Queueing Theory** (1994, 2nd. ed. 2003)
27 Winkler, **Image Analysis, Random Fields and Dynamic Monte Carlo Methods** (1995, 2nd. ed. 2003)
28 Kalpazidou, **Cycle Representations of Markov Processes** (1995)
29 Elliott/Aggoun/Moore, **Hidden Markov Models: Estimation and Control** (1995)
30 Hernández-Lerma/Lasserre, **Discrete-Time Markov Control Processes** (1995)
31 Devroye/Györfi/Lugosi, **A Probabilistic Theory of Pattern Recognition** (1996)
32 Maitra/Sudderth, **Discrete Gambling and Stochastic Games** (1996)
33 Embrechts/Klüppelberg/Mikosch, **Modelling Extremal Events for Insurance and Finance** (1997, corr. 4th printing 2003)
34 Duflo, **Random Iterative Models** (1997)
35 Kushner/Yin, **Stochastic Approximation Algorithms and Applications** (1997)
36 Musiela/Rutkowski, **Martingale Methods in Financial Modelling** (1997, 2nd. ed. 2005)
37 Yin, **Continuous-Time Markov Chains and Applications** (1998)
38 Dembo/Zeitouni, **Large Deviations Techniques and Applications** (1998)
39 Karatzas, **Methods of Mathematical Finance** (1998)
40 Fayolle/Iasnogorodski/Malyshev, **Random Walks in the Quarter-Plane** (1999)
41 Aven/Jensen, **Stochastic Models in Reliability** (1999)
42 Hernandez-Lerma/Lasserre, **Further Topics on Discrete-Time Markov Control Processes** (1999)
43 Yong/Zhou, **Stochastic Controls. Hamiltonian Systems and HJB Equations** (1999)
44 Serfozo, **Introduction to Stochastic Networks** (1999)
45 Steele, **Stochastic Calculus and Financial Applications** (2001)
46 Chen/Yao, **Fundamentals of Queuing Networks: Performance, Asymptotics, and Optimization** (2001)
47 Kushner, **Heavy Traffic Analysis of Controlled Queueing and Communications Networks** (2001)
48 Fernholz, **Stochastic Portfolio Theory** (2002)
49 Kabanov/Pergamenshchikov, **Two-Scale Stochastic Systems** (2003)
50 Han, **Information-Spectrum Methods in Information Theory** (2003)

(continued after index)

The mentioned MITRE repositories of attacks and vulnerabilities, for example, offer lists of typical kinds of consequences (such as loss of integrity or confidentiality), and estimates of typical severity (such as low or high). Others, such as the OWASP list of the top 10 security risks [63], come with estimates of the severity of the technical impact of the attacks. Still, when using such predefined estimates, we always need to adjust them to the specific target, assets, and party in question. Other means to aid the consequence estimation include, for example, security testing such as penetration testing and software testing. This helps risk assessors to judge the severity of vulnerabilities, and to explore the possible outcomes of attacks.

We can use similar sources and techniques for the estimation of likelihoods. In some cases we may be able to estimate the likelihoods of incidents directly, but we often need to analyze the causes of risks, namely the cyber-threats and vulnerabilities. Another advantage of doing the latter is that we get a better understanding of the most important causes of cyber-risk. Such an understanding is useful in particular during the identification of risk treatments.

For the analysis of malicious threats we may use techniques for threat modeling to describe aspects such as attack prerequisites, attacker skills or knowledge required, resources required, attacker motive, attack opportunity, and so forth [8, 51, 61]. Similar descriptions can be made for vulnerabilities, such as ease of discovery and ease of exploit. In combination, this information can be used to derive likelihoods of threats and incidents, as discussed further in Chap. 11.

In analyzing threats and vulnerabilities we make use of the techniques mentioned before. For both malicious and non-malicious threats, the aim is to estimate the likelihood of cyber-threats to occur and the severity of the vulnerabilities that the threats may exploit. In combination, these estimates serve as a basis for achieving the main goal of the risk assessment, namely to estimate risk levels. Knowledge about who or what the threat sources are, how they cause threats, and which vulnerabilities the threats exploit also facilitates the estimation of the consequences of the incidents.

5.3.5 Evaluation of Cyber-risk

Following the risk evaluation as described in Sect. 2.4.4, there are four tasks involved in this step, namely risk consolidation, risk evaluation, risk aggregation, and risk grouping. In the following we explain the particular concerns for each of these tasks for the domain of cyber-risk.

- *Consolidation of risk analysis results:* The consolidation of cyber-risk is similar to the general case; we focus on the cyber-risks for which the estimates are uncertain and where this uncertainty may affect the risk level or our decision making. What is specific to cyber-risk is the distinction between malicious and non-malicious cyber-risk, and we must take care and check for any risks that are both malicious and non-malicious. When estimating such risks, we need to take into account both the malicious threats and non-malicious threats together. Consider, for example, the incident of unauthorized access to some sensitive data. If this incident may be caused by either a hacker or some accidental information leakage, we must ensure that we add up the respective likelihoods.

- *Evaluation of risk level:* The evaluation of cyber-risk is similar to the general case. However, for our own convenience we may choose to evaluate malicious and non-malicious cyber-risks separately.
- *Risk aggregation:* For the risk aggregation we do as in the general case and look for situations in which there are individual risks that must be evaluated together when this may yield a higher combined risk level.
- *Risk grouping:* The grouping of cyber-risk is similar to the general case, apart from one thing. Due to the distinction between malicious and non-malicious cyber-risk, we have this additional and useful way of grouping the identified risks. Some treatments, such as intrusion detection, apply mostly to malicious risks, while other treatments, such as security training, apply mostly to non-malicious risks.

5.3.6 Treatment of Cyber-risk

There are two features in particular that distinguish the risk treatment of cyber-systems from the general case. First, the highly technical nature of cyber-systems means that to a large extent the options for risk treatment are also technical. In addition we need to consider the sociotechnical aspects and human involvement. Second, the distinction between malicious and non-malicious cyber-risks has implications for how we can most adequately treat the risks. In the following we discuss these aspects in relation to the treatment options of risk reduction, risk retention, risk avoidance, and risk sharing.

For the treatment option of risk reduction we seek means to eliminate threat sources and threats, reduce the severity of vulnerabilities, or by other means reduce the likelihood or consequence of incidents. In general, the kinds of means and controls useful for risk reduction include: correction, elimination, prevention, impact minimization, deterrence, detection, recovery, monitoring, and awareness [32]. In order to determine how to most effectively and efficiently reduce risk, we make use of the obtained risk models since they give information about the most likely threats and the most severe vulnerabilities.

For malicious threats it may be hard, if not impossible, to eliminate the threat sources. In some cases it might be possible to bring charges or take legal action, but often we seek other means. To reduce the likelihood of threats and the severity of vulnerabilities, we consider the various parts and aspects of the cyber-system in question, and how it interacts with the cyberspace. This includes applications, servers, clients and networks. For non-malicious threats it may, for example, be possible to eliminate threat sources by implementing technical barriers, such as stricter access control to reduce the chance of accidental leakage of sensitive data. Treatments of a more sociotechnical nature include increased security awareness and training, improved security policies, and improved processes and routines.

When conducting the treatment identification we take into account the cost of the possible means of risk reduction. This includes the cost of acquisition, implementation, administration, operation, monitoring, and maintenance of the treatments. Other aspects to consider are, for example, performance issues and the end-user perspective. Some security mechanisms come at the cost of performance, and we need to make sure that the system continues to fulfill any performance requirements. Usability is of course also important. For end users some security controls are too cumbersome. For example, if a password regime is complex, end users may be inclined to have the credentials written down in clear text, for instance, on a sticker glued to the technical device in question.

In conducting the treatment identification, we make use of techniques such as interviews, brainstorming and questionnaires, focusing on the cyber-threats, vulnerabilities, and incidents that cause unacceptable risks. We may also make use of open lists and databases, such as the ISO standards on ICT security [33, 34].

The treatment option of risk retention is similar for cyber-systems as for the general case. For the option of risk avoidance it is sometimes relevant to look for alternatives to current solutions in case these are exposed to unacceptable cyber-risks. This can, for example, be to terminate the use of cloud services or web applications and replace them with in-house solutions.

The final kind of treatment option is risk sharing, for example by sub-contracting or insurance. A specific kind of insurance that is emerging within the domain of cyber-systems is that of cyber-insurance [15]. Cyber-insurance is the transfer of financial risk associated with network and computer incidents to a third party [6]. The cyber-insurance products and market are still immature, but insurance companies are increasingly offering such policies, in particular in the USA, but also in Europe. There are several challenges related to cyber-insurance, such as the assessment of cybersecurity and cyber-risk in terms of monetary cost and benefit. For some organizations, cyber-insurance may, however, be a good way to reduce their exposure to cyber-risk or to reduce uncertainty regarding cyber-risk.

5.4 Monitoring and Review of Cyber-risk

The process of monitoring and review as described in Sect. 2.5 makes a clear distinction between

- monitoring and review of risk, and
- monitoring and review of risk management

In the first case we are concerned with the system in question; in the second case we focus on the implementation and operation of the risk management process for the system in question. This distinction is of course also relevant within cyber-risk management.

5.4.1 Monitoring and Review of Cyber-risk

We benefit from the fact that cyber-systems are computerized, at least to a large degree. The options for monitoring and surveying cyber-risks are numerous. We can, for example, keep logs of the number and frequency of detected attacks or viruses, monitor the network traffic to detect irregularities, gather information from firewalls and intrusion detection systems. For the purpose of risk monitoring it is useful to identify and specify a set of cyber-risk indicators to be monitored. Such indicators may be the frequency of detected attacks, the frequency of successful attacks, the accumulated downtime of specific services over a given time period, or the frequency of rejected logins due to invalid credentials. The current values of indicators give implicit information about the current risk picture at any point in time. To make the best use of indicators, organizations may define procedures or functions for combining them and for mapping them to explicit risk information.

In order to maximize the value of the cyber-risk information that we gather by system monitoring, risk assessments and open repositories, we need efficient and useful means for representing the information. One option is to establish a classification and categorization of information as mentioned previously. An additional option is to establish a risk register where the information is available to all relevant stakeholders. The register may include, for example, top incidents, threats, and vulnerabilities that stakeholders need to be aware of. How organizations represent the data should be adapted to the user roles, so that, for example, management staff, security personnel, software developers, and system architects get the right kind of information for their individual needs.

5.4.2 Monitoring and Review of Cyber-risk Management

Since cyberspace is a continuously evolving and fast-changing environment, the process of cyber-risk management is required to be more dynamic than a conventional risk management process. In fact, it must be largely computerized, and in the future in almost real time.

This means, of course, that the monitoring and review of cyber-risk management must to the extent possible also be computerized; otherwise it will not be possible to react in time. Hence, cyber-risk managers should aim for a computer-based infrastructure to monitor the performance of the cyber-risk management process itself. This includes not only how risk assessments are planned and conducted, but also how and to what extent measures and controls are implemented and how information is obtained and communicated.

5.5 Further Reading

For up-to-date information about cyber-threats, vulnerabilities, and incidents there are several open lists and repositories that can be used such as the MITRE attack patterns [51] and vulnerability lists [52], as well as the lists of security risks. Such overviews often come with estimates of attack likelihood, vulnerability severity, and incident consequence. There are also several organizations that regularly publish statistics on cyber-incidents and top cyber-risks, such as [61, 66, 73, 76, 82].

Some standards and guidelines on ICT security offer lists of threat sources, threats, and vulnerabilities that can be used as input to the cyber-risk identification. This includes, for example, ISO 27005 [32], ISO 27032 [28] and NIST SP 800-30 [54]. The same kinds of standards and guidelines often offer advice on options for cyber-risk treatment. There is also literature and guidance on attacker modeling, for example as provided by OWASP [64] or the Common Criteria [8].

We also refer to Part II of this book where we demonstrate the whole process of cyber-risk assessment.

Part II
Cyber-risk Assessment Exemplified

Chapter 6
Context Establishment

The objective of Part II is to demonstrate cyber-risk assessment, as presented in Sects. 2.4 and 5.3, by means of a running example. The example concerns an AMI (advanced metering infrastructure) in a smart grid. The example has been made up in order to demonstrate cyber-risk assessment in a way that is easy to understand also for readers not familiar with such infrastructure.

The first step of the cyber-risk assessment is context establishment. Establishing and documenting the context is an essential part of cyber-risk assessment. The outcome of this step guides the rest of the risk assessment; hence, this has a major impact on the success of the overall risk assessment. This chapter gives an example-driven walk-through of the context establishment. Notice that in Part II, to save space we often drop the "cyber"-prefix and write, for example, "risk" although we actually address cyber-risk.

6.1 Context, Goals, and Objectives

A *smart grid* is an electricity distribution network that can monitor the flow of electricity within itself and automatically adjust to changing conditions. A central element in realizing a smart grid is the introduction of an AMI. Such an infrastructure consists of power meters that use two-way communication to collect information related to electric power usage from electricity customers and also to provide information to these customers.

Our example concerns risks for such an infrastructure, which includes components for switching on/off power to an electricity customer or limiting the amount of power provided (choking). National authorities all over the world are pushing for the introduction of smart grid technology. Distribution system operators, which are the organizations responsible for providing electrical power to end users, therefore need to understand what this implies in terms of risk. We will present a risk assessment on behalf of one such distribution system operator. This means that the distribution system operator is the *party* for the risk assessment.

© The Author(s) 2015   51
A. Refsdal et al., *Cyber-Risk Management*, SpringerBriefs
in Computer Science, DOI 10.1007/978-3-319-23570-7_6

6.1.1 External Context

A smart grid is a cyber-physical system that is part of a critical infrastructure. Blackouts can have large societal consequences. The distribution system operator (our party) is therefore subject to a number of national laws and regulations governing its operations, including risk management. As part of the identification of the external context, it is important to identify and document these laws and regulations, although the results are not included here. Failure to comply may have significant legal and financial consequences, in the worst case putting the operator out of business. Moreover, power outages or incidents such as charging the wrong amount or disclosing electricity customer data can damage the reputation and public trust in the operator.

6.1.2 Internal Context

The distribution system operator constituting our party does business by distributing electrical power to electricity customers. The overall goals of the operator are to provide power in a reliable manner so that the electricity customers do not experience unintended power failures, to exchange correct and timely information with customers at all times so that they can be charged the right amount, and to protect the privacy of its customers. Most of the employees of the distribution system operator have strong technical competence and a few of the staff have received special training in risk assessment.

6.1.3 Goals and Objectives

The primary goal of the assessment is to help reduce the risk of incidents that may impact the business of the distribution system operator, by identifying appropriate treatments for the important risks. The secondary goals are to comply with laws and regulations concerning risk management and to be able to document this compliance. Moreover, the distribution system operator wishes the risk assessment to be documented in a way that can be understood by a wide range of internal and external stakeholders, including those who are not themselves experts on cybersecurity or smart grids. Technical details should therefore be avoided as far as possible.

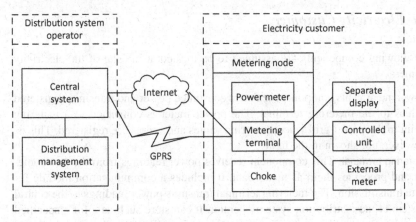

Fig. 6.1 Overview of the target of assessment

6.2 Target of Assessment

Figure 6.1 provides an overall view of the AMI that constitutes our target of assessment. This high-level view, which is adapted and simplified from descriptions found in [45, 79], is sufficient for the purpose of our demonstration. Notice, however, that if our aim was to perform a more detailed assessment, we would most likely include a description of the specific services, communication protocols, encryption mechanisms, physical connection points, and other technical details of the target. Depending on the purpose and scope, we could also include human users and the ways in which they interact with the system, such as user interfaces and operating procedures.

We have chosen to employ an informal notation for capturing the target. Other notations such as UML or domain-specific languages may also be used depending on the type of system and available documentation, as well as the preferences and competence of the risk assessors and the target group.

The system is divided into two main parts enclosed by broken lines and representing different physical locations. Electricity customer represents the end user of power, such as an institution or a private home. Distribution system operator represents our party, the company that owns and maintains the power grid and is responsible for providing power to the electricity customer. Although the figure only shows one electricity customer, there will be many customers for the same distribution system operator.

Components belonging to Electricity customer or Distribution system operator are shown as boxes with solid lines, while communication paths are shown as lines between components. In the following we explain the components in more detail, as well as the communication channels between them.

6.2.1 Electricity Customer

The following components are assumed to be located at the site of the electricity
customer:

- Power meter: This component registers consumed power and transmits registered
  values to the metering terminal. The power meter is connected to a metering
  point in the power grid where metering values and events are registered. This is,
  however, not shown in the figure.
- Metering terminal: This component receives power readings from the power meter and processes them as meter data. It includes a communication module for
  data transmission. The metering terminal transmits power readings to the central
  system at the distribution system operator. It can store such data for a specified
  period of time until it receives confirmation of reception from the central system.
- Choke: This component has functionality for switching the power to the electricity customer on or off, or limiting the power provided. Such functionality can be
  used, for example, to restrict the amount of power provided to individual customers in the case of shortages or lack of payment.
- Metering node: This is a composite component consisting of Power meter, Metering
  terminal, and Choke.
- Separate display: This is a display connected to the metering node for presenting
  information related to power consumption, prices, and cost.
- Controlled unit: This refers to equipment for operating the local installations or
  appliances of the electricity customer based on data received from the metering node. For example, power-consuming appliances such as water heaters and
  washing machines may be set up to turn on or off depending on the current tariff
  and the amount of power currently being consumed by other home appliances.
- External meter: This refers to external meters that can be connected to the metering node, such as meters for gas, heating from a central heating plant, water, and
  so on. For example, this allows the control of an electrical stove (as a controlled
  unit) to depend on data received from a central heating plant.

6.2.2 Distribution System Operator

The following components are assumed to be located at the distribution system operator on whose behalf we conduct the risk assessment:

- Central system: This is the central computer system in the AMI, which collects
  meter data and events from all metering nodes. It also transmits information to
  various ICT systems of the distribution system operator.
- Distribution management system: This is the operational system for the low-
  voltage grid, accessing real-time data and providing information on a single con-
  sole at the control center in an integrated manner.

6.2.3 Communication Channels Between Components

As illustrated in Fig. 6.1, the communication between the electricity customer and distribution system operator may go either via the Internet or GPRS (General Packet Radio Service). This communication is handled by Central system on the Distribution system operator side and by Metering terminal on the Electricity customer side.

At the Electricity customer location, all communication between Metering node and the external components (Separate display, Controlled unit, and External meter) goes to/from Metering terminal via either cable or a local wireless network. Hence, access to these communication paths can only be obtained from the location of Electricity customer.

At the Distribution system operator location, the communication between Central system and Distribution management system goes via cable.

6.3 Interface to Cyberspace and Attack Surface

The infrastructure documented in Fig. 6.1 is part of a larger cyberspace involving the Internet and the mobile network on which the GPRS runs. The interface between the target of assessment and the cyberspace therefore consists of the following:

- The connection point between Central system and the Internet.
- The connection point between Central system and the mobile network.
- The connection point between Metering terminal and the Internet.
- The connection point between Metering terminal and the mobile network.

Attacks can be launched remotely targeting each of these connection points; hence, they are all included in the attack surface. In addition, attacks can be launched locally on all the components and local communication lines. For example, an attacker close to the Electricity customer location may use Metering terminal to get access to the system either by physically plugging into the device or by breaking into the local wireless network used for communication between Metering terminal and the external components.

Table 6.1 identifies the attack surface, distinguishing between Electricity customer and Distribution system operator. The left-hand column shows whether the attack requires remote or local access. Notice that the latter includes, for example, side-channel attacks based on electromagnetic leaks, which requires the attacker (and/or her equipment) to be physically close to the target, but not necessarily within the premises of the distribution system operator or the electricity customer.

Table 6.1 Attack surface

Attacker location	Electricity customer	Distribution system operator
Attack from remote location (worldwide)	Connection point between Metering terminal and either the Internet or GPRS	Connection point between Central system and either the Internet or GPRS
Attack from physical location close to target	Metering node; External components (Separate display, Controlled unit, and External meter); Communication line between Metering node and each of the external components	Central system; Distribution management system; Communication line (cable) between Central system and Distribution management system

6.4 Scope, Focus, and Assumptions

In order to ensure that the time and resources available for the assessment are spent on the most important matters, we need to be clear about the scope, focus, and assumptions of our assessment. By properly documenting these elements we also make sure that this essential information is available for users of the assessment results, regardless of whether they were involved in the assessment or not.

6.4.1 Scope

We limit the scope of the assessment to risks due to attacks on or via the target of assessment as described in Sect. 6.2. This means, for example, that attacks via back-end systems that may be connected to Central system but are not included in Fig. 6.1, such as an electricity customer database, are outside the scope of the assessment. Communication between Central system and Metering node via GPRS is supposed to be subject to a separate assessment and is therefore also outside our scope.

6.4.2 Focus

The focus of the assessment is first and foremost on the exchange of meter data and control data via the Internet and the ways in which this may affect the provisioning of power, as the distribution system operator is particularly concerned about this aspect. Although within the scope, the main focus will not be on attacks via physical access to components.

Risks caused by malicious as well as non-malicious threat sources should be considered. Regarding functionality, the focus of the assessment is on basic AMI functions, which include registering electricity customer meter data, transfer of data

Table 6.2 Assets

Asset	Explanation
Integrity of meter data	The integrity of meter data should be protected all the way from Power meter to Distribution system operator
Availability of meter data	Meter data from Metering node should be available for Distribution system operator at all times
Provisioning of power to electricity customers	Power should only be switched off or choked as a result of legitimate control signals from Central system

between Electricity customer and Distribution system operator, and switching on/off or choking of power provided to the electricity customer.

6.4.3 Assumptions

We assume that threat sources may be internal as well as external. This applies to malicious as well as non-malicious threats. Furthermore, as the target of assessment is part of a critical infrastructure, we assume that it may be targeted not only by individuals with a purely financial or personal motive, but also by actors who wish to disrupt society. Finally, we assume that all meter data and control data sent between the central system and metering nodes are encrypted.

6.5 Assets, Scales, and Risk Evaluation Criteria

The final part of the context establishment step consists of identifying assets and defining scales and risk evaluation criteria. By identifying the assets of the party in question, namely the distribution system operator, we make it clear what we need to protect, which is essential to determine what risks are relevant. The scales determine how to measure risk, while the risk evaluation criteria define the risk levels.

6.5.1 Assets

The distribution system operator is the sole party for the cyber-risk assessment, which means that all consequence assessments and risk evaluation criteria will be defined from that perspective. Table 6.2 shows the identified assets. Notice that, for example, privacy issues, such as the confidentiality of meter data, are not considered here.

Table 6.3 Likelihood scale

Likelihood value	Description	Definition
Rare	Less than once per ten years	$[0,1\rangle : 10y = [0,0.1\rangle : 1y$
Unlikely	Less than once per two years	$[1,5\rangle : 10y = [0.1,0.5\rangle : 1y$
Possible	Less than twice per year	$[5,20\rangle : 10y = [0.5,2\rangle : 1y$
Likely	Two to five times per year	$[20,50\rangle : 10y = [2,5\rangle : 1y$
Certain	Five times or more per year	$[50,\infty\rangle : 10y = [5,\infty\rangle : 1y$

6.5.2 Likelihood Scale

Risk levels will be determined from the likelihood and consequence of incidents. We therefore need to define suitable scales. For this risk assessment we specify likelihood in terms of frequencies. Table 6.3 presents the likelihood scale.

In general, the granularity of the chosen scales depends on availability of data and the preferences of the decision makers. For this assessment we use five-step scales of intervals for likelihood as well as consequence values. Hence, during the risk analysis we need only determine which interval the likelihood of an incident lies within, rather than providing an exact value. Such use of intervals is a simple way of expressing the uncertainty that is typically involved when making these kinds of estimates. In Chap. 13 we discuss how to deal with uncertainty in further detail.

6.5.3 Consequence Scales

As consequence descriptions will differ between assets, we create a separate table for each of them. Tables 6.4, 6.5 and 6.6 show the consequence descriptions for integrity of meter data, availability of meter data, and provisioning of power, respectively. Remember that the scales are defined from the perspective of the distribution system operator, which considers the overall business impact of potential incidents. Clearly, the scales would look quite different if defined, for example, from the perspective of an electricity customer.

Notice that the descriptions of consequence values are meant only as an aid to assess the degree of damage for identified incidents. For example, if we estimate that the consequence of a given incident in terms of harm to the availability of meter data is comparable to meter data for 1,001-10,000 electricity customers being unavailable for one day (24 hours), then we should assign the consequence value "Minor" to this incident.

Table 6.4 Consequence scale for integrity of meter data

Consequence value	Description
Insignificant	Errors in meter data for up to 100 electricity customers
Minor	Errors in meter data for 101-2,000 electricity customers
Moderate	Errors in meter data for 2,001-20,000 electricity customers
Major	Errors in meter data for 20,001-50,000 electricity customers
Critical	Errors in meter data for more than 50,000 electricity customers

Table 6.5 Consequence scale for availability of meter data

Consequence value	Description
Insignificant	Meter data for up to 1,000 electricity customers unavailable for 1-24 hours
Minor	Meter data for up to 1,000 electricity customers unavailable for more than 1 day or meter data for 1,001-10,000 electricity customers unavailable for 1-24 hours
Moderate	Meter data for 1,001-10,000 electricity customers unavailable for more than 1 day or meter data for more than 10,000 electricity customers unavailable for 1-24 hours
Major	Meter data for more than 10,000 electricity customers unavailable for 25 hours-7 days
Critical	Meter data for more than 10,000 electricity customers unavailable for more than 7 days

Table 6.6 Consequence scale for provisioning of power to electricity customers

Consequence value	Description
Insignificant	Power outage for up to 100 electricity customers for 1-24 hours
Minor	Power outage for up to 100 electricity customers for more than 24 hours or power outage for 101-1,000 electricity customers for 1-24 hours
Moderate	Power outage for 101-1,000 electricity customers for more than 24 hours or power outage for 1,001-10,000 electricity customers for 1-24 hours
Major	Power outage for 1,001-10,000 electricity customers for 25-72 hours or power outage for more than 10,000 electricity customers for 1-24 hours
Critical	Power outage for 1,001-10,000 electricity customers for more than 72 hours or power outage for more than 10,000 electricity customers for more than 24 hours

6.5.4 Risk Evaluation Criteria

We are now ready to define the risk evaluation criteria. Figure 6.2 shows the risk matrix that defines our criteria. In this case we have chosen the same risk matrix for all assets. Three risk levels have been defined, represented by light grey (*Low* risk level), medium grey (*Medium* risk level) and dark grey (*High* risk level). The risk levels help guide the identification and selection of treatments. For high risks we always identify potential treatments. For low and (in particular) medium risks, we should ideally also identify treatments, but this is less critical and treatments have lower priority than for high risks. In the end, the decision on which treatments to implement will of course be taken based on an evaluation of the cost versus the expected risk reduction.

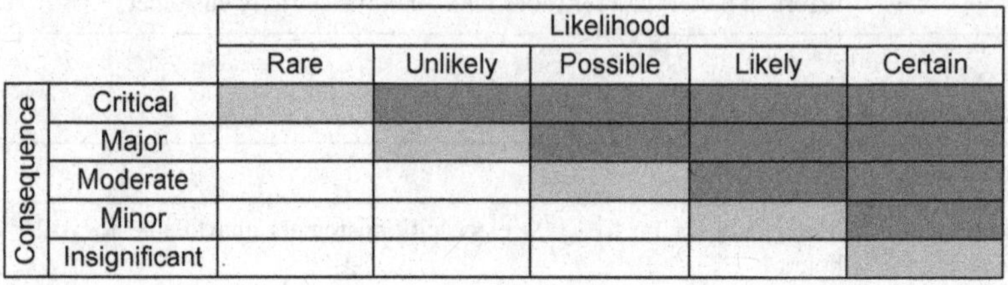

		Likelihood				
		Rare	Unlikely	Possible	Likely	Certain
Consequence	Critical					
	Major					
	Moderate					
	Minor					
	Insignificant					

Fig. 6.2 Risk matrix

6.6 Further Reading

The NIST guidelines for smart grid cybersecurity [56] gives a high-level view of a smart grid architecture focusing on logical interfaces, as well as high-level security requirements.

Chapter 7
Risk Identification

After establishing the context, we are ready to start identifying risks. Here the goal is to arrive at a collection of threat sources, threats, vulnerabilities, incidents, and risks that is as correct and complete as possible for our particular target of assessment and assets. We start by giving an overview of some risk identification techniques, before moving on to identification of risks caused by malicious threats, as described in Sect. 5.3.2, and risks caused by non-malicious threats, as described in Sect. 5.3.3.

7.1 Risk Identification Techniques

Since cyber-systems are computer based, there is normally a lot of data and information available from event logs, intrusion detection systems and other monitoring tools, vulnerability scanners, results from penetration tests or other kinds of security tests, source code reviews, and so on. When identifying risk we try to fully exploit such information. Therefore we perform a systematic walk-through of the target description, including the attack surface and assets, in order to identify any such information sources to be used. These sources are mapped to the relevant part(s) of the target, which will also be useful in the risk analysis step later. Typically, this is done in close cooperation with maintenance personnel, technical managers, security managers, or others who have detailed knowledge about the technical infrastructure. For example, any test results concerning the metering terminal interface to the Internet are mapped to this particular part of the attack surface. These test results then help us to identify vulnerabilities and threats for attacks through this interface. Table 7.1 illustrates a simple way of documenting this kind of information. The first column shows which part of the target system, attack surface, or asset the information relates to. The second column briefly describes the kind of information and source, while the third column provides a reference to the source.

Notice that, when using historical data such as event logs, you should take great care not to fall into the trap of believing that tomorrow will be like yesterday. Even if a certain threat has not materialized in the past, it does not mean that it cannot do

© The Author(s) 2015   61
A. Refsdal et al., *Cyber-Risk Management*, SpringerBriefs
in Computer Science, DOI 10.1007/978-3-319-23570-7_7

Table 7.1 Results from tests, monitoring logs, and so on of relevance to risk identification

Part of target / asset	Source description	Reference
Connection point between metering terminal and external meters	A test of the metering terminal interface to external meters was performed last year. The test included checking whether there is adequate input sanitation. A written report documents the test procedure and results.	MeterTest.docx
Availability of meter data	The central system logs all instances of meter data from the metering node of an electricity customer not being received at the expected point in time. The logs for the last three-year period have been compiled in a single pdf file.	MissingMeterData.pdf

so in the future. The absence of corresponding events from the logs does not mean that a threat or incident should be left out of the assessment. This is particularly important to remember with respect to rare incidents with a high consequence, such as a large-scale coordinated attack on the metering infrastructure. Similarly, even if a vulnerability is not detected by a security test, it does not mean that it does not exist. For the risk identification we need not consider the severity of vulnerabilities or the likelihood of threats and incidents; at this point we document everything that may be relevant and leave the further analysis for later.

Throughout the risk identification, and also during the risk analysis later, we make sure to carefully consider whether there are parts of the target for which more security testing, logging/monitoring, or other probing is needed. This is, however, also a question of available time and resources. Furthermore, it depends on whether the required information can be obtained by other means.

In addition to the target-specific information sources discussed above, valuable input to the risk identification can also be found in open sources such as international standards, online repositories, and various reports on cybersecurity, threats, and vulnerabilities. When exploiting such input, our main challenge is to identify the specific sources of relevance, and to select from these sources only those elements that are relevant to our assessment. Here we recommend a simple four-step approach:

1. Establish relevance criteria based on, for example, the kind of system or domain you are dealing with, the assets, or the risk type.
2. Identify information sources based on the established criteria. For an overview of open sources of information, see Chap. 5. We also give some examples of references regarding specific parts of the risk identification process throughout this chapter.
3. Select from these sources only those elements that are relevant to your assessment.
4. Reformulate the selected elements, which by necessity are described in general terms, so that they apply specifically to your target of assessment and assets.

Even if we are dealing with cyber-systems, it is essential for the risk identification to extract information not only from system logs, security tests, and so on, but also from people who know the target of assessment well from their particular viewpoints. For our assessment, these people may include the developers of the central system or metering nodes, the maintenance team and operators of the central system, the information security officer and managers of the distribution system operator, and potentially also some of their electricity customers.

External experts may also possess valuable knowledge for our assessment; although they do not know the specific target of assessment, they may provide general information about typical threat sources, vulnerability and attack types, and trends. When interacting with external experts you must of course take great care not to disclose confidential information unless this has been approved by the party on whose behalf we conduct the assessment.

For obtaining information from people, we may employ interviews. Interviews can follow a strict structure where all questions are planned in advance, but we can also use an open format with key themes to be covered, yet with considerable openness to additional inputs from the interviewee. The most appropriate option is usually a mixed approach where we prepare questions, but are ready to follow up on any unforeseen but relevant issues that the interviewee brings up. Interviews may provide very valuable information, but must be used with care. Interviews are quite resource intensive and depend on the right persons being willing and available. Carrying out the interviews and compiling and aggregating results also require skill from the risk assessors.

Another option for extracting information and knowledge from people is the use of questionnaires. This is easier to organize than interviews, as we do not have to agree with the subject on a date. On the downside, we lose the possibility of asking follow-up questions or making clarifications. Moreover, the subject has little opportunity to elaborate on issues that are not covered by the questionnaire, meaning that we may lose important information.

We can also make use of brainstorming and similar techniques for risk identification. This involves gathering together relevant stakeholders and personnel with first-hand knowledge about specific parts or aspects of the target to contribute to the identification process in plenary sessions. A big advantage of this approach is that the participants are able to discuss and to follow up on each other's ideas. For example, if one participant identifies a vulnerability not thought of by anyone else, then all of them can think of ways in which threats can exploit this vulnerability. Assuming we are able to gather the right people, this can potentially prove very successful. Unfortunately there are also some pitfalls associated with brainstorming that we need to keep in mind. One is that the personalities of the participants play a major role, and there is a danger that the more outspoken persons dominate while others hardly contribute, so that not all views are brought forward. Individual participants may also take the opportunity to pursue their own agenda and focus only on issues that are within their own area of interest. Other pitfalls are that the discussion can digress off topic and that the available time may not be properly distributed between the topics to be covered.

Successful brainstorming therefore requires a highly skilled risk assessor to lead the sessions. It also requires that we make plans in advance for how to structure and guide the discussions. The structure can be based on, for example, assets, threat source types, vulnerability types, or parts of the target description or attack surface. How we choose to structure the brainstorming is up to us, but in general it depends on the target of assessment, any preferences of the participants, and which step of the risk identification we are dealing with. How to do on-the-fly documentation of the proceedings may also pose a challenge. We therefore need to appoint a dedicated secretary with this responsibility. If all participants consent, we could of course use video or audio recordings, but we do not generally recommend this, as it is likely to inhibit the participants. On the more practical side, gathering together all the participants for a brainstorming session may also be difficult.

Which information sources and techniques to use for the risk identification depends on a number of factors, such as available resources and information sources, and the kind of target. For example, for a standard web application or service of a non-critical system, a satisfactory risk identification can probably be based to a large degree on generic standards and repositories of cyber-threats and vulnerabilities. On the other hand, when dealing with a highly specialized critical system such as the AMI, the risk identification is much more involved. We therefore seek to combine techniques to get as complete a picture as possible and to confirm the results. For example, if interviews reveal uncertainty about the presence of certain kinds of vulnerabilities or the feasibility of attacks, then vulnerability scanning and security testing can help to reduce the uncertainty.

For documenting and structuring risk assessment results, in Part II of this book we employ tables and textual descriptions. We consider tables to be well suited for our purposes since the assessment will be done at a generic level, without going deeply into the technical details of how threats and incidents materialize. Common alternatives to tables are various kinds of graph-based risk-modeling techniques. For an overview of risk-modeling techniques and their area of use, see ISO/IEC 31010 [30].

7.2 Malicious Risks

As explained in Sect. 5.3.2, when identifying malicious risks we basically need to understand how a game between an adversary and the defender may play out. How the adversary may launch attacks, which vulnerabilities he or she may exploit, and what incidents may result if the attack succeeds depend on who the adversary is. Therefore we start by identifying relevant adversaries, which we refer to as threat sources. Then we move on to threat identification, where we describe potential attacks with respect to the assets in question, before identifying vulnerabilities, and finally incidents.

Notice that although this order offers a good way to structure the identification process, it only serves as a guideline. We are free to deviate whenever it serves the

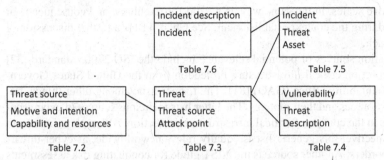

Fig. 7.1 Overview of tables documenting risks caused by malicious threats

overall goal and to go back and update previous results at any time. For example, if a constructive discussion about vulnerabilities starts during the threat source identification, then we make sure to document all relevant comments, and go back to the threat sources later. The important thing is to establish a collection of relevant threat sources, threats, vulnerabilities, incidents, and risks that is both consistent and as complete as possible. The results we present in this chapter show the final outcome of the identification process.

Figure 7.1 gives an overview of the tables that we use to document risks caused by malicious threats, as well as the relations between these tables. Each box represents a table. The uppermost compartment shows the main column, the heading of which occurs in boldface in the actual table. The lowermost compartment represents the rest of the columns. Lines between tables indicate entries that occur in more than one table. For example, threats occur also in the vulnerability table since vulnerabilities are considered in relation to threats. The table number is indicated below each box. As we will explain in the following, the tables are designed to accommodate the risk identification approach. The risks may be deduced implicitly. For each pair of an incident and asset harmed by the incident there is one risk.

7.2.1 Threat Source Identification

To identify malicious threat sources we need to understand who may want to initiate attacks and why. For this purpose we consider all possible motives and intentions, including financial gain, revenge or grudges, political or religious agendas, espionage, or simply fun and a desire to prove one's ability. The potential for causing harm will to a large degree depend on the motive and intention of the threat sources, as well as their capabilities and available resources. It is therefore important to document these characteristics in the threat source descriptions, and we have designed Table 7.2 accordingly. For this assessment we have chosen to use free text to capture threat source characteristics. This will provide valuable input to the analysis of likelihood and consequence later. Alternatively, we could have defined quantitative or qualitative scales, in the same way as we did for likelihood and consequence of incidents during the context establishment. See Chap. 11 for a further discussion of this approach.

Information sources of potential relevance include the ISO 27005 standard [32] and the report on critical infrastructure protection from the United States Government Accountability Office (GAO) [81]. The former lists, among other things, human threat sources and their motives. The latter lists malicious sources of cybersecurity threats in the context of critical infrastructure protection. Although written from a US perspective, this generic list is equally relevant worldwide, as cyber-threats know no borders. Another source is the NIST guide for conducting risk assessments [54], which lists a number of malicious threat sources.

Table 7.2 documents the malicious threat sources that we identified for the assessment of the smart grid AMI based on the target and the gathered data. Notice that although the descriptions in the table are quite generic, we have selected each of them because they are of relevance to our specific target of assessment, as this is a potential target for their attacks. Understanding how these threat sources can cause concrete threats is the task of the subsequent identification of malicious threats. Before moving on to that we explain the reasoning behind the inclusion of some of the documented threat sources to illustrate the approach.

Script kiddie: Attacks on power supply systems may potentially get a lot of media attention, as power supply concerns everyone. This applies not only to the provisioning of power, but also to corresponding billing and payment services. Such systems may therefore be attractive targets for script kiddies seeking attention.

Cyber-terrorist: A power supply system is a critical infrastructure. Blackouts and disruptions can have huge societal consequences in any modern society. Power supply systems are therefore highly attractive targets for cyber-terrorists seeking to disrupt society or cause societal crises or emergencies.

Black hat hacker: The billing and payment of electric power involve high economic values. A black hat hacker able to tamper with power consumption data could for example use this ability to offer an "electric power bill reduction service" on the black market.

7.2.2 Threat Identification

For each malicious threat source we identify the threats it may initiate. Table 7.3, which documents malicious threats, therefore includes a *Threat source* column as well as the *Threat* column. In addition, for this task we focus specifically on how the threat sources may exploit the attack surface identified during the context establishment. Therefore we also include a separate *Attack point* column to show which parts of the attack surface are being exploited by each threat. By including these three elements in the table format, we also document the explanation behind the identified threats. This is necessary both for the later risk analysis and for the final reporting of the risk assessment results.

Table 7.2 Malicious threat sources

Threat source	Motive and intention	Capability and resources
Script kiddie	Achieve status among a group or prove his/her ability to cause harm. Will seldom be very persistent if faced with difficulties and initial failure	Relatively unskilled, unable to perform complicated attacks. Typically uses tools developed by others to initiate attacks. Very limited access to computational or monetary resources
Cyber-terrorist	Cause disruption in a society through cyber-attacks, preferably against critical infrastructure. Strong political, ideological, or religious motives and willingness to go to extremes	May command significant resources and skill, in some cases even being supported by nation states. Able to perform long-term planning, preparation, and carrying out of attacks
Black hat hacker	Motivated by personal gain, for example through tampering with data or blackmail. This includes, for example, electricity customers who seek to reduce their electricity bill by tampering with meter data	The skill level of black hat hackers can vary a lot, but the best are world-leading experts on cybersecurity. If part of a larger criminal organization, they can also command significant resources
Hacktivist	Similarly to cyber-terrorists, hacktivists are motivated by a political, ideological, or religious agenda and use cyber-attacks to achieve their goals. Although the distinction between cyber-terrorists and hacktivists is fuzzy at best, we assume that hacktivists are less willing to go to extremes and that their aim is to harm selected groups, politicians, or other individuals, rather than society as a whole	Skill level and resources can vary a lot. Most hacktivists are assumed to operate alone or in small or poorly organized groups. However, if well organized they can potentially have access to significant computational resources as well as competence
Insider	An insider is a disloyal employee or consultant of the distribution system operator who is typically motivated either by personal gain or by a desire to harm the employer due to conflicts and discontent	May have access to all systems and possess detailed information and knowledge about the system architecture, functionality, and security features
Malware	By malware we mean here malicious software developed to harm computerized systems, but which are not aimed specifically at harming the assets of the party of the risk assessment	Developers of malware are often highly skilled. Malware can therefore cause significant harm to systems based on standard off-the-shelf operating systems or other software

Again, the examples of typical threats provided by ISO 27005 offer useful input for the threat identification. Other examples include the section on attack mechanisms in ISO 27032 [28], the attack vector descriptions provided by the OWASP top 10 [63], and the representative examples of malicious threats found in the NIST guide for conducting risk assessments [54]. CAPEC [51] also offers an online database of cyber-attack patterns.

Descriptions in sources such as the above are, of course, not specific to our target of assessment. We therefore make sure to describe each of the relevant threats as it applies to our particular target. Table 7.3 documents the results, some of which we explain further below.

DDoS attack on the central system: Regarding script kiddies we consider DDoS attacks as a potential threat, as DDoS attacks have been well known for a long time and a lot of information about how to launch such attacks is available online. It is also possible to buy services for such attacks on the black market. DDoS attacks are also a relevant threat with respect to cyber-terrorists, who may launch such attacks in an attempt to disrupt power provisioning.

Tampering with all or most control data in transit from the central system to the choke component: This threat can be initiated by a cyber-terrorist attack in order to disrupt the power supply. As control data from the central system can be used to choke or disconnect power for electricity customers, tampering with such data for a large number of customers can cause significant societal disruption, which can be a goal for cyber-terrorists. The control data are sent over the Internet. Tampering with data in transit therefore represents a relevant threat for our assessment.

Malware to manipulate meter data is installed on the metering terminal through connection to the external meter: Black hat hackers can potentially make a profit from manipulation of meter data. One way to achieve this is to install malware on the metering terminal, so that manipulated data are sent to the central system. The metering terminal is often connected to external meters, and such connections represent a potential attack point for installation of malware.

7.2.3 Vulnerability Identification

For each malicious threat we identify the existing vulnerabilities that the threat may exploit. Table 7.4, which documents these findings, therefore includes a *Threat* column as well as the *Vulnerabilities* column. We also include a *Description* column, which allows us to provide a more extensive description of the vulnerability. During the identification we pay special attention to the attack point documented as part of the threat identification, as well as any weaknesses of defense mechanisms, or lack of such mechanisms.

With respect to exploiting external information sources, a full chapter of the NISTIR 7628 guidelines for smart grid cybersecurity [53] is dedicated to listing vulnerabilities, divided into four classes: 1) people, policy, and procedure; 2) platform software/firmware vulnerabilities; 3) platform vulnerabilities; and 4) network. ISO 27005 offers a list of vulnerabilities related to hardware, software, network, personnel, site, and organization. Other sources of general vulnerabilities include the online resources offered by OWASP [61] and the common weakness enumeration offered by MITRE [52].

In addition to exploiting such sources, we also take advantage of the fact that our target of assessment is an executing system that can be subject to vulnerability

Table 7.3 Malicious threats

Threat source	Attack point	**Threat**
Script kiddie	Internet connection to the central system	DDoS attack on the central system
Cyber-terrorist	Same as the row above	Same as the row above
Cyber-terrorist	Internet connection between the central system and the metering terminal	Tampering with all or most control data in transit from the central system to the choke component
Black hat hacker	Internet connection between the central system and the metering terminal	Tampering with data in transit from the metering terminal to the central system
Black hat hacker	Communication line between the metering terminal and the external meter	Malware to manipulate meter data is installed on the metering terminal through connection to the external meter
Malware	Internet connection to the metering terminal	Metering node infected by malware
Hacktivist	Internet connection between the metering terminal and the central system	Tampering with control data in transit from the central system to the choke components for selected electricity customers
Insider	Central system	Illegitimate control data sent to the choke components from the central system

scanning and other forms of security testing. This helps us to check whether suspected vulnerabilities are actually present, and may also reveal new vulnerabilities. The use of tests for identification of vulnerabilities should of course be documented. This could, for example, be done by including a separate column with a reference to related tests and test results for each vulnerability. We have chosen not to include such a column here, as going further into the actual tests would be beyond our scope. Table 7.4 documents the results of the identification of vulnerabilities with respect to malicious threats.

Inadequate attack detection and response on central system: Successful protection against DDoS attacks requires firstly that the system is able to detect the attack and secondly that an adequate defense can be initiated. The attack detection mechanism on the central system may, however, be outdated. It may not be clear whether it is able to catch the more advanced kinds of DDoS attacks we have seen in recent years; hence, this needs further investigation. The response is based purely on dropping packets according to fixed classification rules and gives little opportunity for analysing the attack.

Weak encryption and integrity check: Concerning all the identified threats involving tampering with data in transit, members of the central system maintenance team had concerns with the encryption strength of meter data and control data. Hence, this is a potential vulnerability that needs further investigation. The same applies to the integrity checking of such data.

Unprotected local network, no sanitation of input data from the external meter: With respect to the threat of malware being installed on the metering terminal through the connection to the external meter, a test of this interface indicated that there was no sanitation of input to the metering terminal from external meters, which leaves this component vulnerable to injection attacks. This is particularly worrying since the distribution system operator has no means to ensure the protection of the electricity customer's local network over which the metering terminal communicates with the external meter.

Notice that some approaches prescribe the identification of controls or barriers, which are means to prevent threats occurring and/or leading to incidents. For our assessment we consider this to be covered by the vulnerability identification, in the sense that a weak or nonexistent control/barrier constitutes a vulnerability. For example, encryption can be used as a barrier against confidentiality breaches. Weak encryption, for example due to a weak cryptographic hash function or poor protection of keys, will therefore constitute a vulnerability.

7.2.4 Incident Identification

Before moving on to the risk analysis step we need to identify the incidents that may result from the threats and actually harm our assets. In other words, we need to think of potential ways that our assets can be harmed by the threats. Table 7.5, which documents incidents, therefore contains a *Threat* column and an *Asset* column in addition to the *Incident* column. Vulnerabilities are of course also considered, but have not been included in this table, as they are already related to the threats in Table 7.4. Notice that Tables 7.2, 7.3, 7.4, and 7.5 together provide information about each complete chain consisting of a threat source, threat, vulnerability, incident, and affected asset. As mentioned above, for each pair of an incident and asset harmed by the incident there is a risk.

The general information sources we consult to identify incidents are to a large degree the same as we used for threats and vulnerabilities. For example, the list of threats provided by ISO 27005 [32] also includes information about incidents that may result from the threats, such as unauthorized system access and system tampering. OWASP [63] descriptions include information about results of attacks, such as denial of access, user sessions being hijacked, and so on. The attack pattern enumeration offered by MITRE [51] also provides similar descriptions. As in the case of vulnerabilities, security testing can also be used to gain a better understanding of potential incidents.

When exploiting general information sources such as those mentioned above, we make sure to tailor the descriptions so that every incident is clearly expressed in a way that relates specifically to our particular target of assessment and assets. Table 7.5 documents the results of the identification of incidents resulting from malicious threats. The descriptions are given at a high level of abstraction, which is in

Table 7.4 Vulnerabilities with respect to malicious threats

Threat	**Vulnerability**	Description
DDoS attack on the central system	Inadequate attack detection and response on central system	New forms of DDoS attacks are continuously being developed to defeat existing countermeasures. Due to the challenges of keeping the central system running 24/7, combined with the lack of a strong tradition for cybersecurity awareness in the power distribution domain (which has not traditionally operated in cyberspace), countermeasures to various forms of DDoS attacks on the central system are rarely updated and may therefore be out of date
Tampering with all or most control data in transit from the central system to the choke component	Weak encryption and integrity check	The encryption of messages between the central system and the metering node may be weak compared to the current standard. The same applies to the integrity checking of received messages. This applies in particular at the metering nodes, which have relatively little computing power and are rarely replaced
Tampering with data in transit from the metering terminal to the central system	Weak encryption and integrity check	The considerations here are the same as in the previous row
Tampering with control data in transit from the central system to the choke components for selected electricity customers	Weak encryption and integrity check	The considerations here are the same as in the previous row
Malware to manipulate meter data is installed on the metering terminal through connection to the external meter	Unprotected local network, no sanitation of input data from the external meter	The local network at the electricity customer location cannot be assumed to be properly protected, as this depends on the individual customer. Moreover, data from the external meter to the metering terminal are not adequately sanitized before further processing, thereby leaving the metering terminal vulnerable to code injection attacks
Metering node infected by malware	Outdated antivirus protection on metering node	The metering node is connected to the Internet in order to communicate with the central system and is therefore susceptible to malware. However, the virus protection on the metering node is rarely updated
Illegitimate control data sent to the choke components from the central system	Four-eyes principle not implemented, no logging of actions of individual central system operators	The operating procedures and technical implementation of the central system do not enforce approval of control data by a second authorized person. An operator is therefore able to send control data that are not legitimate. Moreover, there is no logging of the actions of individual operators

Table 7.5 Incidents caused by malicious threats

Threat	Incident	Asset
DDoS attack on the central system	Data from metering nodes cannot be received by the central system due to DDoS attack	Availability of meter data
Tampering with all or most control data in transit from the central system to the choke component	False control data received by all or most choke components	Provisioning of power to electricity customers
Tampering with data in transit from the metering terminal to the central system	False meter data for a limited number of electricity customers received by the central system	Integrity of meter data
Malware to manipulate meter data is installed on the metering terminal through connection to the external meter	Same as the row above	Same as the row above
Metering node infected by malware	Malware compromises meter data	Integrity of meter data
Metering node infected by malware	Malware disrupts transmission of meter data	Availability of meter data
Metering node infected by malware	Malware disrupts the choke functionality	Provisioning of power to electricity customers
Tampering with control data in transit from the central system to the choke components for selected electricity customers	False control data received by the choke components for selected electricity customers	Provisioning of power to electricity customers
Illegitimate control data sent to the choke components from the central system	Power supply to electricity customers is switched off without legitimate reason	Provisioning of power to electricity customers

line with the directions given by the distribution system operator during the context establishment. Below we explain the reasoning behind some of the entries.

Data from metering nodes cannot be received by the central system due to DDoS attack: This incident may result from a DDoS attack on the central system, since a successful attack will keep this system too busy serving illegitimate requests. This means that meter data becomes unavailable to the distribution system operator, at least temporarily.

False control data received by all or most choke components: Clearly, tampering with all or most control data in transit from the central system to the choke component may lead to this incident, as the control data will not constitute authentic data sent from the central system. Since the threat source in this case is a cyber-terrorist, it is likely that the control data will be manipulated so as to disrupt the provisioning of power to the electricity customers.

False meter data for a limited number of electricity customers received by the central system: This incident may result from tampering with data in transit from the metering terminal to the central system, which is a threat initiated by a black

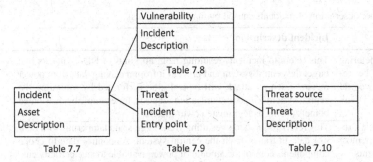

Fig. 7.2 Overview of tables documenting risks caused by non-malicious threats

hat hacker. Reception of false meter data would of course harm the integrity of the meter data.

Often it is useful to provide more information about identified incidents than what is offered by Table 7.5. We therefore include further descriptions in Table 7.6.

7.3 Non-malicious Risks

As explained in Sect. 5.3.3, for risks where no malicious intent is involved, we start from the assets in order to guide the identification process and ensure that we maintain the right scope. The first step is to identify accidental incidents that may harm the assets. Only threats, vulnerabilities, and threat sources that relate to such incidents are relevant. Starting with identification of incidents therefore helps us to focus the rest of the process on the important elements. Having identified non-malicious incidents, the next steps are to identify the weaknesses of the target that make the incidents possible, that is to say the vulnerabilities, and the threats that may lead to the incidents. Finally, we identify the threat sources that can initiate these threats. Similarly to the case of malicious risks, the above order provides a useful guideline for structuring the identification, although we allow ourselves to deviate from this order whenever appropriate. The results we present show the final outcome of the process.

Figure 7.2 gives an overview of the tables that we use to document risks caused by non-malicious threats. They are designed to support the identification process outlined above, but are sufficiently flexible to also accommodate other approaches.

Table 7.6 Further description of incidents caused by malicious threats

Incident	Incident description
Data from metering nodes cannot be received by the central system due to DDoS attack	This refers to incidents resulting from all kinds of DDoS attacks that target the central system and prevent it from receiving data from power meters. Such an attack will typically be performed by saturating the central system with communication requests, for example by distributed botnets posing as legitimate power meters
False control data received by all or most choke components	This refers to incidents resulting from large-scale tampering with control data in transit from the central system. As control data to choke components control the amount of power available to an electricity customer, such threats can lead to widespread brownouts or even blackouts. In order to succeed in sending false messages, an attacker must get into the communication path, intercept and modify legitimate messages or create new messages, and ensure that that the modified or new messages are considered valid by the choke components. However, if the default action of these components is to switch off in the absence of valid control data, then a blackout can be achieved simply by preventing legitimate messages from reaching the choke components
False meter data for a limited number of electricity customers received by the central system	This refers to such incidents resulting from either tampering with meter data in transit from metering terminals or malware on the metering terminals. The technical ways in which the former may be achieved is similar to the previous case. However, as data from metering nodes to the central system primarily concern power consumption, the motive would most likely be financial gain. A black hat hacker could for example offer to manipulate data in order to reduce electricity bills for a suitable fee
Malware compromises meter data	Metering terminals connected to the Internet may be infected by malware even if no malicious person has physical access, thereby affecting their ability to correctly register meter data
Malware disrupts transmission of meter data	Similarly to the case above, malware may affect the ability of the metering terminal to correctly transmit meter data to the central system
Malware disrupts the choke functionality	As the choke component receives control data from the central system via the metering terminal, malware on the metering terminal may prevent it from forwarding correct control data to the choke component. It is also possible that the choke component itself is infected by malware via the metering node
False control data received by the choke components for selected electricity customers	This refers to the same kind of incident as described for *False control data received by all or most choke components* above, except that only a small group is targeted. This would require attackers to be able to identify the specific control signals going to the target group, but is otherwise similar to the above case
Power supply to electricity customers is switched off without legitimate reason	This refers to cases where an insider does this on purpose. This would be a breach of operating procedures and require that the insider knows how to operate the control signals from the central system, but would not otherwise require any specialized cybersecurity or programming skills

7.3.1 Incident Identification

In order to identify incidents caused by non-malicious threats we start from the assets by considering how these can be harmed. Table 7.7, which documents the identified incidents, therefore includes an *Asset* column in addition to the *Incident* column. We also include a *Description* column which allows us to provide further explanation of each incident.

When identifying incidents we pay special attention to the way the assets relate to or are represented in the system. With respect to integrity and availability of meter data, we notice that an electricity customer's power consumption is read by the power meter and fed to the metering terminal, which transmits meter data to the central system over the Internet. Incidents can therefore affect these assets all along this chain. For provisioning of power to electricity customers, control data are sent from the central system to the metering terminal and forwarded to the choke component. Depending on the received data, this component may switch off or reduce the amount of power provided to the customer. To aid the identification of incidents, we make use of sources such as system logs, monitored data, repositories of previous incidents or other historical data, and input from people with knowledge about the target system. Table 7.7 documents incidents caused by non-malicious threats. In the following we explain the reasoning behind the inclusion of some of the rows to illustrate the approach.

Communication between the central system and the metering terminal is lost: If this communication is broken then the choke component will not receive control data from the central system, which disrupts the provisioning of power. In addition, the central system will not receive meter data from the metering terminal.

Software bug on the metering terminal compromises meter data: Transmission of correct meter data depends on the correct functioning of the metering terminal. A software bug may cause a malfunction that can potentially result in corrupted meter data being sent to the central system, thereby compromising the integrity of these data.

Mistakes during maintenance of the central system disrupt transmission of control data to the choke component: Provisioning of power may be disrupted if correct control data are not received by the choke component. This means that misconfiguration of communication parameters or other maintenance mistakes that disrupt transmission of control data may also disrupt power provisioning.

The metering terminal goes down due to damage from lightning: If the metering terminal goes down then it will not be able to transmit meter data or receive control data. This incident will therefore harm availability of meter data and provisioning of power to the affected electricity customers.

Table 7.7 Incidents caused by non-malicious threats

Asset	Incident	Description
Provisioning of power to electricity customers; Availability of meter data	Communication between the central system and the metering terminal is lost	Provisioning of power to the electricity customer depends on control data being sent from the central system to the metering terminal. Availability of meter data depends on such data being sent in the opposite direction
Integrity of meter data	Software bug on the metering terminal compromises meter data	Metering terminals run software to register meter data and transmit these to the central system. Software bugs on metering terminals may therefore compromise meter data
Availability of meter data	Software bug on the metering terminal disrupts transmission of meter data	Similarly to the above case, software bugs on metering terminals may disrupt transmission of meter data to the central system
Provisioning of power to electricity customers	Software bug on the metering terminal disrupts the choke functionality	Control signals to the choke component from the central system go via the metering terminal. Software bugs on metering terminals may therefore disrupt the choke functionality by not forwarding correct control signals
Provisioning of power to electricity customers	Mistakes during maintenance of the central system disrupt transmission of control data to the choke component	Maintenance mistakes such as misconfiguration of communication parameters may prevent or disrupt transmission of control data
Availability of meter data	Mistakes during maintenance of the central system prevent reception of data from metering nodes	Maintenance mistakes such as misconfiguration of communication parameters may prevent metering node data from being received
Provisioning of power to electricity customers; Availability of meter data	The metering terminal goes down due to damage from lightning	Lightning may result in physical damage to the metering terminal which prevents it from functioning

7.3.2 Vulnerability Identification

For each identified incident we look for vulnerabilities that allow the incident to occur or that increase its likelihood. Table 7.8, which documents vulnerabilities with respect to non-malicious incidents, therefore includes an *Incident* column in addition to the *Vulnerability* column. We also include a *Description* column to provide more information about the vulnerability.

Vulnerabilities with respect to non-malicious threats are often related to the ability of operators or other persons interacting with the system to perform their tasks as expected. We therefore pay special attention to human, social, and organizational factors such as training, skills, time pressure, and procedures. For our assessment this applies, for example, to those who operate and maintain the central system. Similar considerations also apply to suppliers on whose services or products the system depends. We also consider the technical vulnerabilities of software, hardware, and other equipment that affect our target of assessment. The sources of vulnerability descriptions to be found in the literature are largely the same as for the malicious case, although the relevant entries may of course differ.

Table 7.8 documents the results of the identification of vulnerabilities with respect to non-malicious threats. The reasoning is explained below.

Single communication channel between central system and metering terminal: Many electricity customers do not have the possibility of communication via GPRS and rely solely on the Internet connection. This is an obvious weakness with respect to maintaining communication.

Poor testing: This vulnerability applies to the software of metering terminals, which run quite complicated software. This software is responsible for registering power readings from the power meter and transforming these readings into meter data to be submitted to the central system. It is also responsible for receiving control data from the central system and forwarding these to the choke component. In addition, there are the general protocols and functionality for communication with the central system and external components such as the controlled unit. Extensive testing according to state-of-the-art methods is therefore required.

Poor training and heavy workload: This applies to members of the maintenance team responsible for the central system, which is the single most important component of our target of assessment. Maintenance of the central system is very difficult, as it consists of a number of hardware and software components, communicates with other systems, and needs to run continuously. A log of previous errors raises doubts about whether all members of the maintenance team have the required skills and experience. The experienced members of the maintenance team have a very heavy workload and may not always be available when needed.

Inadequate overvoltage protection: Metering terminals include computing hardware that is not very robust with respect to transient overvoltages, for example caused by lightning. It is doubtful whether the overvoltage protection of most electricity customers provides the required level of protection.

7.3.3 Threat Identification

Having identified vulnerabilities for each incident, we move on to identify the threats that may, due to the vulnerabilities, cause the incidents to occur. Each threat is related to (at least) one incident and corresponding vulnerability. To link the threats to incidents, we include an *Incident* column as well as a *Threat* column in Table 7.9, which documents non-malicious threats. Moreover, we go through the target description to find parts or components where threats may occur. To document this we also include an *Entry point* column. Table 7.9 documents the results of the non-malicious threat identification.

Internet connection to the metering terminal goes down: Given the vulnerability of a single communication channel between the central system and the metering

Table 7.8 Vulnerabilities with respect to non-malicious threats

Incident	Vulnerability	Description
Communication between the central system and the metering terminal is lost	Single communication channel between central system and metering terminal	For many electricity customers there is no redundant communication link to the central system
Software bug on the metering terminal compromises meter data	Poor testing	The software for the metering terminals is developed and tested by the metering terminal supplier. Previous experience indicates that their testing routines are not satisfactory
Software bug on the metering terminal disrupts transmission of meter data	Same as the row above	Same as the row above
Software bug on the metering terminal disrupts the choke functionality	Same as the row above	Same as the row above
Mistakes during maintenance of the central system disrupt transmission of control data to the choke component	Poor training and heavy workload	Maintenance of the central system is highly challenging due to its complexity and the need to operate 24/7. Hence, performing these tasks requires extensive training and experience. The persons that have the required skills also have a heavy workload, meaning that less qualified personnel sometimes need to carry out the tasks
Mistakes during maintenance of the central system prevent reception of data from metering nodes	Same as the row above	Same as the row above
The metering terminal goes down due to damage from lightning	Inadequate overvoltage protection	Robust overvoltage protection is needed to protect the metering terminals from lightning

terminal for many electricity customers, it is clear that communication will be lost if the metering terminal loses its Internet connection.

Buggy software distributed on metering terminals: We have identified three incidents involving software bugs, and poor testing has been shown to be a vulnerability. Distribution of buggy software is therefore an important threat to consider.

Mistakes during update/maintenance of the central system: Two of our incidents concern mistakes during update/maintenance of the central system. Such mistakes therefore constitute a relevant threat, in particular in the light of poor training and heavy workload having been identified as a vulnerability.

Table 7.9 Non-malicious threats

Incident	**Threat**	Entry point
Communication between the central system and the metering terminal is lost	Internet connection to the metering terminal goes down	Internet connection to the metering terminal
Software bug on the metering terminal compromises meter data	Buggy software distributed on metering terminals	Metering terminal
Software bug on the metering terminal disrupts transmission of meter data	Same as the row above	Metering terminal
Software bug on the metering terminal disrupts the choke functionality	Same as the row above	Metering terminal
Mistakes during maintenance of the central system disrupt transmission of control data to the choke component	Mistakes during update/maintenance of the central system	Central system
Mistakes during maintenance of the central system prevent reception of data from metering nodes	Same as the row above	Central system
The metering terminal goes down due to damage from lightning	Electricity customer home/building is struck by lightning	Metering terminal

7.3.4 Threat Source Identification

It now remains to identify threat sources. For each threat we ask what its potential source can be. Table 7.10, which documents non-malicious threat sources, therefore includes a *Threat* column as well as the *Threat source* column. An additional *Description* column lets us provide extra information about the threat source.

When identifying non-malicious threat sources we focus on technical errors occurring in the target of assessment or in systems on which it depends, persons that may make mistakes or behave in unforeseen ways when legitimately interacting with the target, and natural phenomena such as lightning and flood. As an aid in this task, ISO 27005 [32] provides a nice overview of potential threat sources, divided into categories such as physical damage, natural events, and technical failures. In addition, NIST [54] provides lists of non-malicious threat sources divided into the categories accidental, structural, and environmental sources. Potential threat sources that we need to consider for our assessment include, for example, those who operate and maintain the central system, software and hardware components on the side of the distribution system operator and the electricity customer, and all communication links. Table 7.10 shows the result of the identification of non-malicious threat sources.

Table 7.10 Non-malicious threat sources

Threat	Threat source	Description
Internet connection to the metering terminal goes down	Internet connection to the metering terminal	Problems with the connection may initiate threats to the communication between the metering terminal and central system
Buggy software distributed on metering terminals	Software bug	Any kind of software error or malfunction that arises due to mistakes rather than malicious intent
Mistakes during update/maintenance of the central system	Maintenance personnel	Persons responsible for maintaining the computer systems and infrastructure for the distribution system operator. They do not seek to cause harm, but may still do so by mistake, neglect, or lack of proper training. Notice that a maintenance person with malicious intent is considered to be an insider with respect to this risk assessment
Electricity customer home/building is struck by lightning	Lightning	Strokes of lightning which may have potential for causing damage to computerized systems and network infrastructure

7.4 Further Reading

An overview of vulnerabilities for smart grids can be found in the guidelines for smart grid cybersecurity from NIST [53]. In their recommendations for protecting industrial control systems, ENISA gives a high-level view of the current situation of technological threats with respect to protecting such systems [14]. EUROPOL [16] provides an assessment of Internet organized-crime threats from a European police perspective. The document includes a section on vulnerabilities of critical infrastructure that specifically addresses smart grids. Although threats are described at a very generic level, documents like this can add a useful perspective to the risk identification. For an overview of general threats, vulnerabilities, and other information not specifically addressing smart grids or other critical infrastructures, we refer to Sect. 5.5. With respect to combining risk analysis and testing, the OWASP testing guide [64] provides some discussion of the relationship between security testing and risk analysis. There is also an emerging field of research in this area that will hopefully mature further in the near future [10, 11, 21].

Chapter 8
Risk Analysis

Having identified cyber-threats, vulnerabilities, and the incidents that constitute risks by harming the identified assets, our next task is to assess the likelihood of these incidents and their consequence for each of the affected assets, so that the risk level can be determined. In order to achieve this, it is usually necessary to perform an analysis of the related threats and vulnerabilities. This also helps us to better understand what contributes to the risk, which is useful for identifying treatments.

As illustrated by Fig. 8.1, we analyze the likelihood for threats to materialize in terms of the frequency scale defined during the context establishment. For severity of vulnerabilities we prefer to use a simple scale consisting of the steps *High*, *Medium*, and *Low*, as the severity cannot be directly captured by frequencies. Finally we analyze the likelihoods and consequences of incidents in terms of the scales defined during the context establishment. The information sources we exploit for the risk analysis are more or less the same as those used for risk identification. The main difference is that now we also need to consider the severity of vulnerabilities and the likelihood of threats and incidents, as well as the consequence of incidents, rather than simply determining whether the threats, vulnerabilities, and incidents are relevant or not. In the following we demonstrate the reasoning behind the analysis for some selected examples. When documenting the risk assessment we make sure to include all the information sources and reasoning behind the analysis, typically in an appendix to the risk assessment report.

8.1 Threat Analysis

We start the risk analysis by analyzing threats. Due to their different nature, it is normally useful to look at malicious and non-malicious threats separately, which is what we do below. However, the distinction is not always clear-cut, and some threats may be both malicious and non-malicious. In such a case, we usually prefer to include it among the malicious threats. However, the important thing is that the threat is not left out and that both its malicious and non-malicious aspects are considered.

© The Author(s) 2015
A. Refsdal et al., *Cyber-Risk Management*, SpringerBriefs
in Computer Science, DOI 10.1007/978-3-319-23570-7_8

Fig. 8.1 Overview of risk
analysis process for our smart
grid risk assessment

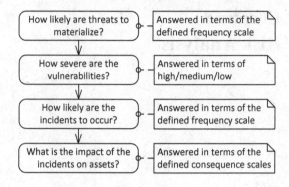

8.1.1 Malicious Threats

We start by analyzing the threat *DDoS attack on the central system* from Table 7.3.
Our main sources of information in estimating its likelihood are the event logs provided by the distribution system operator, as well as the expert judgments of the
participants. We choose to follow an approach inspired by the OWASP risk-rating
method [62], which uses the threat source factors of skill level, motive, opportunity,
and size to analyze the threat. The factors are rated on a scale from 0 to 9.

As the first step in analyzing the threat of a DDoS attack, we consider the threat
source. In this case we have actually identified two different threat sources, a script
kiddie and a cyber-terrorist. We therefore look at each of them in turn.

Table 7.2 tells us that the script kiddie is relatively unskilled and unable to perform complicated attacks. We therefore assign skill rating 3. We have no reason to
believe that script kiddies have specific interest in this particular power distribution
system, but we know they are sometimes attracted to critical infrastructures in general. The motive for conducting a specific attack is also generally weak, so for this
we assign rating 1. The opportunity is a measure of the resources and opportunities
that the script kiddie requires to conduct the attack, for which we assign rating 7. Finally, size is a measure of how large this group of threat sources is. As script kiddies
can reside anywhere in the world, we assign the rating 7.

For the cyber-terrorist, we again consult Table 7.2. Based on the description there,
we assign skill level 7, motive 8, and opportunity 7. Since we assume that the number of cyber-terrorists is much lower than that of script kiddies, we assign a size
rating of 3, even if cyber-terrorists may also reside anywhere in the world.

The OWASP method prescribes taking the average of the threat source factors
to obtain an overall rating for the threat. For the script kiddie, the average equals
4.5. For the cyber-terrorist, the average equals 6.25. This means that the threat level
is quite high. Based on these results and using our own likelihood scale defined
in Table 6.3, we estimate the likelihood of this threat to be *Likely*, as documented
in Table 8.1. At this point, you should always check that the estimate is supported
by the available event logs and confirmed by the participants from the distribution
system operator. We follow a similar approach for the remaining malicious threats.

Table 8.1 Malicious threat analysis

Threat	Likelihood	Estimate basis/comments
DDoS attack on the central system	Likely	This kind of attack is frequent and requires modest skills and resources. The estimate is confirmed by event logs and by cybersecurity statistics
Tampering with all or most control data in transit from the central system to the choke component	Possible	Dedicated cyber-terrorists are probably able to perform such attacks. Although few instances have been recorded so far, reports from Interpol and similar agencies give reason to expect increased likelihood of this type of attack
Tampering with data in transit from the metering terminal to the central system	Possible	Complaints from electricity customers and subsequent investigations indicate that this kind of tampering has occurred in recent years, although the number of incidents and affected customers is hard to estimate
Malware to manipulate meter data is installed on the metering terminal through connection to the external meter	Possible	Tests have revealed that such an attack is technically not very difficult, but it requires access to the connection to the external meter. Although the number of incidents may be fairly high, the number of affected meters will therefore be small
Metering node infected by malware	Rare	Although the metering node is connected to the Internet, it is quite different from standard computers, and does not run most of the software targeted by most malware
Tampering with control data in transit from the central system to the choke components for selected electricity customers	Unlikely	Such threats have not yet been observed by the central system operator. However, recent developments may indicate that activists are increasingly willing to target political adversaries in this way
Illegitimate control data sent to the choke components from the central system	Unlikely	The central system operator has not experienced any such instances, but this has happened in other companies. Employee satisfaction surveys indicate that the central system operator employees are loyal. However, if an insider actually wants to perform such an attack, she is likely to succeed

8.1.2 Non-malicious Threats

Non-malicious threats are by nature unexpected or unintended events that happen by accident or by chance. In analyzing such threats we also start by considering the threat source. By understanding who or what may cause the threat, we can better understand how likely the threat is to occur. Further sources of information include event logs, expert judgments, interviews or questionnaires, and available statistics about the typical likelihood of similar threats in enterprises and other organizations. For the analysis of non-malicious threats we need to keep in mind that threats and near incidents that have occurred before may not be reported or registered. This can, for example, be due to lack of reporting routines and due to the reluctance of personnel to do self-reporting. To establish a better basis for the analysis of non-malicious threats we can investigate relevant properties of the organization, such as culture, routines, skills, security awareness, procedures, and so forth.

For the threat *Mistakes during update/maintenance of the central system* from Table 7.9, for example, we first consider the threat source, namely the *Maintenance personnel*. By identifying who they are and what their responsibilities and job tasks are, we get an understanding of how and when they can cause the identified threat. In addition to this we may base our estimate on the event log and on the judgment made by, for example, the head of the maintenance team. There is evidence that mistakes are made on an almost monthly basis on average, as documented in Table 8.2 by the estimate *Certain*. Notice that at this point we estimate the likelihood of any mistake, no matter how grave. We follow a similar approach for the remaining non-malicious threats.

Table 8.2 Non-malicious threat analysis

Threat	Likelihood	Estimate basis/comments
Internet connection to the metering terminal goes down	Certain	This includes cases where individual electricity customer's homes lose Internet connection, which according to general statistics happens very often
Buggy software distributed on metering terminals	Possible	This estimate is based on patching logs for various software products developed by the provider of metering terminal software during the last four years
Mistakes during update/maintenance of the central system	Certain	This estimate is based on event logs and statements from the head of the management team
Electricity customer home/building is struck by lightning	Certain	This estimate is based on statistics for the geographical area where the electricity customers are located

8.2 Vulnerability Analysis

The next step consists of analyzing vulnerabilities. For this we choose to use a simple scale consisting of the steps *High*, *Medium*, and *Low*. Again we distinguish between vulnerabilities with respect to malicious and non-malicious threats.

8.2.1 Malicious Threat Vulnerabilities

For the assessment of the severity of the identified vulnerabilities we can again make use of the information sources of expert judgments, statistics, and open repositories. But we can also investigate our target of assessment by conducting, for example, vulnerability scans, security testing, penetration testing, and code review.

The identified vulnerability regarding the DDoS attack according to Table 7.4 is *Inadequate attack detection and response on central system.* Inspired by the OWASP risk-rating method we rate vulnerabilities by using the factors of ease of discovery, ease of exploit, awareness, and intrusion detection.

The ease of discovery is a measure of how easy it is for the possible threat sources to come upon this vulnerability. Checking whether systems are vulnerable to DDoS attacks is often straightforward, and we therefore assign the value 7 (easy) to this factor. For the ease of exploit we need to investigate the target of assessment, and perhaps conduct some testing. Already during the risk identification we ascertained that this is a vulnerability that obviously can be exploited. Security testing can confirm this by demonstrating that most illegitimate requests are indeed not detected. For this reason we assign the rating 5 (easy) for the ease of exploit factor. The awareness factor is a measure of how well known this vulnerability is to the threat sources in question. As the knowledge of the existence of such vulnerabilities is widespread we assign the value 6 (obvious) to this factor. Similarly to the ease of exploit, we have already established that the intrusion detection is rather weak, partly based on insights from the experts and partly based on results from security testing. We assign the value 7 to this factor, meaning that intrusions are usually not detected when they happen. This leaves us with an average vulnerability score of 6.25. We consider 6.25 out of 9 to be quite severe, and therefore assign severity *High*, as documented in Table 8.3. We complete the table following a similar approach.

8.2.2 Non-malicious Threat Vulnerabilities

For the non-malicious threats there is of course no intent to discover and exploit vulnerabilities. Instead we seek to understand the extent to which there is a lack of barriers that could prevent threats from leading to incidents.

For the incidents resulting from the threat *Mistakes during update/maintenance of the central system*, for example, we identified the vulnerability *Poor training and heavy workload*, as shown by Table 7.8. After investigating the background and expertise of the maintenance personnel, looking into the tasks and routines, and interacting with the head of the maintenance team we may for example establish that the staff has strong training and expertise in system development. The security awareness, however, may be somewhat weak among some of the personnel, and we may also find that the workload during some periods is very heavy, at least for key personnel. At the same time, the routines for reviewing updates and testing the system before launching the updates are strong and thorough. Overall, our estimate

Table 8.3 Vulnerability analysis with respect to malicious threats

Vulnerability	Severity	Explanation
Inadequate attack detection and response on central system	High	Tests revealed that the DDoS attack detection mechanism is unlikely to detect a large part of illegitimate traffic. The response is based purely on dropping packets, which leaves little possibility of analyzing attacks
Weak encryption and integrity check	Medium	Inspections revealed that a weak encryption scheme is used for the data exchanged between the metering terminals and the central system. The same applies to the integrity checking
Unprotected local network, no sanitation of input data from the external meter	Medium	The central system operator has no control of the local networks of electricity customers. We must therefore assume that such networks may be poorly protected. There is no sanitation of input data from external meters to the metering terminals
Outdated antivirus protection on metering node	High	The antivirus protection on metering nodes is rarely updated
Four-eyes principle not implemented, no logging of actions of individual central system operators	High	Inspection of policies and interviews revealed that all tasks can be performed by a single operator. Moreover, the actions of individual operators on the central system are not logged

of the severity of the vulnerability in question is therefore *Medium*, as documented in Table 8.4. The remaining vulnerabilities have been addressed in a similar manner.

8.3 Likelihood of Incidents

In order to obtain an initial estimate of the likelihood of the incidents we consider the analysis of threats that lead to the incidents and the vulnerabilities that the threats exploit.

The incident *Data from metering nodes cannot be received by the central system due to DDoS attack* from Table 7.5, for example, is due to the threat *DDoS attack on the central system* and the vulnerability *Inadequate attack detection and response on central system*. For the threat we assigned likelihood *Likely*. The vulnerability severity was set to *High*, indicating that a large portion of the threat occurrences will actually lead to the incident. Although the number of DDoS attacks that succeed will likely be lower than the number of attempts, we still estimate that the frequency for the incident also lies within the interval of *Likely* on our scale. We retain this estimate even though event logs show only two such incidents for the last three years (which corresponds to *Possible*). This is because the threat and vulnerability analysis, supported by recent reports documenting increasingly advanced DDoS attacks on critical infrastructure, give good reasons to believe the frequency will increase

Table 8.4 Vulnerability analysis with respect to non-malicious threats

Vulnerability	Severity	Explanation
Single communication channel between central system and metering terminal	High	The Internet connection is the only communication channel to the central system for many electricity customers
Poor testing	Medium	Inspection of maintenance logs revealed a number of instances where bugs have been discovered in the metering terminal software. Previous experience indicates that the testing routines of the external software provider are unsatisfactory, and the central system operator does not test software updates for metering terminals before deployment
Poor training and heavy workload	Medium	Interviews indicate that security awareness is not high. Key persons have too much to do. Routines for reviewing and testing updates to the central system before deployment are strong
Inadequate overvoltage protection	High	The computing hardware of metering terminals is not robust with respect to transient overvoltage

compared to previous years. The result is documented in Table 8.5, which includes likelihood as well as consequence estimates for all malicious incidents.

To estimate the likelihood of the incidents that are caused by non-malicious threats, we also make use of the results of the analysis of threats and vulnerabilities. For example, the two incidents *Mistakes during maintenance of the central system disrupt control signals to the choke component* and *Mistakes during maintenance of the central system prevent reception of data from metering nodes* are caused by the same threat, as shown by Table 7.9. We estimated that the likelihood of the threat *Mistakes during update/maintenance of the central system* is *Certain*, and that the severity of the relevant vulnerability, namely *Poor training and heavy workload*, is *Medium*. At first glance this could be taken to imply that the two incidents occur with the same frequency. However, we found before that there are routines in place for reviewing and testing the system before changes are launched. Because provisioning of power to the electricity customer is more critical than the continuous reading of meter data, the routines are stronger with respect to updates and changes that may affect control data. This observation, combined with the data logs, leads us to the likelihood *Unlikely* regarding control data to the choke component, and the likelihood *Possible* regarding the reception of meter data. These estimates, together with the likelihoods and consequences for the other non-malicious incidents, are documented in Table 8.6.

8.4 Consequence of Incidents

The consequence of an incident must be judged for each asset it harms. For the incident *Data from metering nodes cannot be received by the central system due to DDoS attack*, for example, we need to estimate the consequence for the asset *Availability of meter data* according to the scale defined in Table 6.5. Therefore we need to consider the expected time it takes to detect and respond to an attack, as well as the number of affected electricity customers. In the experience of the distribution system operator, which is supported by their internal investigation reports of the incidents, the DDoS attacks that have occurred before have never caused loss of availability for more than one day. The number of electricity customers whose meter data becomes unavailable can, however, be higher than before, as the customer base has increased. Based on this information we therefore assign the consequence estimate *Moderate* to the incident in question, as documented in Table 8.5, which includes the estimates for all incidents resulting from malicious threat sources.

The provisioning of power to electricity customers is more critical than the availability of the meter data. This is also reflected by the consequence scales for the respective assets. This explains the consequence *Moderate* regarding the choke component (risk no. 14) and the consequence *Minor* regarding the meter data (risk no. 15), as documented in Table 8.6, which also includes the likelihood and consequence estimates for the remaining non-malicious incidents.

Table 8.5 Likelihood and consequence for incidents caused by malicious threats

No.	Incident	Asset	Likelihood	Consequence
1	Data from metering nodes cannot be received by the central system due to DDoS attack	Availability of meter data	Likely	Moderate
2	False control data received by all or most choke components	Provisioning of power to electricity customers	Unlikely	Critical
3	False meter data for a limited number of electricity customers received by the central system	Integrity of meter data	Likely	Minor
4	Malware compromises meter data	Integrity of meter data	Rare	Moderate
5	Malware disrupts transmission of meter data	Availability of meter data	Rare	Moderate
6	Malware disrupts the choke functionality	Provisioning of power to electricity customers	Rare	Major
7	False control data received by the choke components for selected electricity customers	Provisioning of power to electricity customers	Rare	Insignificant
8	Power supply to electricity customers is switched off without legitimate reason	Provisioning of power to electricity customers	Unlikely	Moderate

Table 8.6 Likelihood and consequence for incidents caused by non-malicious threats

No.	Incident	Asset	Likelihood	Consequence
9	Communication between the central system and the metering terminal is lost	Provisioning of power to electricity customers	Certain	Minor
10	Same as the row above	Availability of meter data	Certain	Insignificant
11	Software bug on the metering terminal compromises meter data	Integrity of meter data	Unlikely	Moderate
12	Software bug on the metering terminal disrupts transmission of meter data	Availability of meter data	Unlikely	Moderate
13	Software bug on the metering terminal disrupts the choke functionality	Provisioning of power to electricity customers	Rare	Major
14	Mistakes during maintenance of the central system disrupt transmission of control data to the choke component	Provisioning of power to electricity customers	Unlikely	Moderate
15	Mistakes during maintenance of the central system prevent reception of data from metering nodes	Availability of meter data	Possible	Minor
16	The metering terminal goes down due to damage from lightning	Provisioning of power to electricity customers	Likely	Insignificant
17	Same as the row above	Availability of meter data	Likely	Insignificant

8.5 Further Reading

For analyzing threats there are a number of sources and methods available. In addition to those provided by OWASP [62], there is the CAPEC catalogue offered by MITRE [51]. This gives ratings of attack prerequisites, attacker skills or knowledge, required resources, and attack indicator/warning. The CWE catalogue [52] also gives useful input for analysis of vulnerabilities.

Chapter 9
Risk Evaluation

At this point we have identified the risks and analyzed their likelihood and consequence. From this we can establish the risk level and compare it to the risk evaluation criteria, as explained in Sect. 2.4.4 and Sect. 5.3.5. We also need to consider whether some risks that we have regarded as separate are actually instances of the same risk and therefore should be aggregated and evaluated as one risk. Furthermore, as preparation for the risk treatment, we group risks according to relationships such as shared vulnerabilities or threats. However, as analysis of likelihood and consequence is notoriously difficult, we start by reviewing the results from the previous step in order to check whether any adjustments need to be made.

9.1 Consolidation of Risk Analysis Results

The goal of the consolidation of risk analysis results is to make sure that the correct risk level is assigned to each risk. This is important because the risk levels direct the identification of treatments and provide essential decision support for the management. The central question is not whether each likelihood and consequence estimate is correct, but rather whether the resulting risk level is correct. For example, for risk no. 4 in Table 8.5, we assigned likelihood *Rare* and consequence *Moderate*, which according to the risk evaluation criteria defined by Fig. 6.2 gives risk level *Low*. Even if the likelihood is increased to *Unlikely*, the risk level will remain *Low*. Hence, for this risk, the distinction between these two likelihood levels is not essential for determining the risk level. On the other hand, if we are uncertain whether the consequence for risk no. 15 should remain at *Minor* or perhaps be increased to *Moderate*, then we need to investigate the issue, as this would bring the risk level from *Low* to *Medium*. When consolidating analysis results we direct our attention to the risks where 1) we are uncertain about the likelihood and/or consequence estimate and 2) this uncertainty may affect the risk level or the risk treatment.

We also make sure to check whether there are any risks that are both malicious and non-malicious. This is typically the case if malicious and non-malicious threats

© The Author(s) 2015
A. Refsdal et al., *Cyber-Risk Management*, SpringerBriefs
in Computer Science, DOI 10.1007/978-3-319-23570-7_9

can result in the same incident. In our case, this would mean that the same incident occurs in both Table 8.5 and Table 8.6. In such cases we need to check that the likelihood and consequence estimates are consistent, and that both the malicious and the non-malicious causes have been considered when estimating the likelihood. This can be easy to overlook since we are dealing with the malicious and non-malicious risks separately during much of the risk assessment.

As part of the consolidation we also revisit the risk evaluation criteria defined during the context establishment. Sometimes decision makers will want to adjust the criteria based on any new insights gained through the process so far, or on the results of the analysis.

The results of the consolidation are documented in the same place as the risk analysis results simply by making the necessary corrections and updates, and also adding references if new information sources have been used. For our analysis, this would mean updating the relevant entries in the tables presented in Chap. 8.

9.2 Evaluation of Risk Level

Having consolidated the risk analysis results, we are ready to evaluate the risks. The risk level of each risk is determined by its likelihood and consequence according to the risk matrix. In our case, risk evaluation is performed simply by plotting each risk in the risk matrix defined in Fig. 6.2. The result for malicious risks is shown in Fig. 9.1, where the numbers refer to the risk numbers in Table 8.5. Figure 9.2 shows the result for non-malicious risks from Table 8.6.

		Likelihood				
		Rare	Unlikely	Possible	Likely	Certain
Consequence	Critical		2			
	Major	6				
	Moderate	4,5	8		1	
	Minor				3	
	Insignificant	7				

Fig. 9.1 Risk matrix with malicious risks from Table 8.5

9.3 Risk Aggregation

During the evaluation we need to take into account that some risks may "pull in the same direction" to the degree that they should actually be evaluated as a single risk. There are basically two cases where this may hold.

		Likelihood				
		Rare	Unlikely	Possible	Likely	Certain
Consequence	Critical					
	Major	13				
	Moderate		11,12,14			
	Minor			15		9
	Insignificant				16,17	10

Fig. 9.2 Risk matrix with non-malicious risks from Table 8.6

The first case, which is illustrated by Fig. 9.3, concerns incidents that harm more than one asset of the same party, thereby giving rise to more than one risk for the party in question. Even if the risk of incident X harming asset A and the risk of incident X harming asset B are both low, it may be that the combined effect of harm to A and B warrants a higher risk level for the aggregation of these risks. In this case the likelihood of the aggregated risks remains the same, while the consequence is the joint consequence of the two risks.

Fig. 9.3 Aggregation of risks where one incident harms more than one asset of the same party

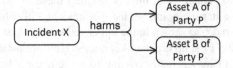

The second case is illustrated by Fig. 9.4 and concerns a single asset being harmed by more than one incident. Even if the risk of each individual incident harming the asset in question is low, it may be that the combined effect on the asset yields a higher risk. A typical situation in which we might aggregate is when the incidents are of the same nature, as is the case for Y_1 and Y_2 in Fig. 9.4 a), or when the occurrences of the incidents are triggered by the same threat, as is the case for U and V in Fig. 9.4 b). Notice that this also needs to be taken into account in cases where one of the incidents is malicious and the other is non-malicious.

Whatever the case and whatever the situation, we need not aggregate unless this can bring the aggregated risk to a new risk level. The risk level is, after all, what matters with respect to decision making. For a set of risks that are acceptable only if considered individually, deciding not to aggregate can give a false impression that no treatments are needed. Such decisions should therefore be taken with care.

We now return to our assessment. Going through Table 8.5 and Table 8.6 we find that there are no instances where a single incident harms more than one asset. Hence, the type of aggregation illustrated by Fig. 9.3 is not relevant for us.

However, risk no. 4, *Malware compromises meter data*, and risk no. 11, *Software bug on the metering terminal compromises meter data*, both concern software on the metering nodes and harm the integrity of meter data. They can therefore be viewed as special instances of a more generic incident, which we can call *Software on the*

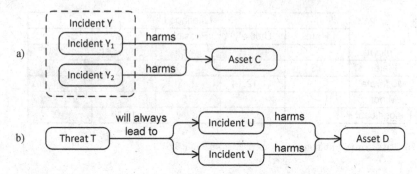

Fig. 9.4 Aggregation of risk where a) two incidents are special instances of a common, more abstract instance, or b) two incidents are triggered by the same threat

metering node compromises meter data. Hence, they are candidates for aggregation as per Fig. 9.4 a). Looking at their risk levels in Figs. 9.1 and 9.2, we notice that their places in the risk matrix give reason to think that aggregation may yield a higher risk level than is given by either of the individual risks. We therefore decide to perform the aggregation. This is done by aggregating likelihood and consequence values separately, and then combining these to obtain the risk level in the usual way. As a starting point, we list the incidents, likelihoods, and consequences of the original risks, as shown in the upper rows of Table 9.1.

First up are the likelihoods. Here we notice that the incidents of risks nos. 4 and 11 may actually overlap to some degree. For example, malware may compromise meter data that are already compromised by a software bug. Moreover, the likelihoods are given as intervals rather than exact values, which means that adding up likelihoods may yield a new interval that spans more than one step of the likelihood scale defined in Table 6.3. This means that we cannot simply sum up the likelihoods of the contributing incidents, but need to use our judgment. After careful considerations about the nature of the incidents and the degree of overlap, we may for example arrive at likelihood *Possible* for the aggregated risk.

Next up are the consequences. Since the aggregated incident represents a generalization of each of the original incidents, rather than a combined occurrence, it clearly would not make sense to add up their consequences. Unless we are considering instances where simultaneous occurrences of several incidents cause additional harm, the consequence of the aggregated incident should not be greater than the highest of the original consequences. A good rule of thumb is that if all the original incidents have the same consequence, then we use the same value for the aggregated incident. If they do not, we can either use some kind of average value, possibly weighted according to likelihoods, or resolve the issue by consulting representatives of the party of the asset. In our case, we notice that risks nos. 4 and 11 both have consequence *Moderate*, hence this is also the value we use for the aggregated risk. The lowermost row of Table 9.1 shows the result. The plus sign denotes aggregation.

Similarly to the above case, it seems reasonable to aggregate risks nos. 5 and 12, and risks nos. 6 and 13. For the rest we decide to retain the original risks. Fig. 9.5

Table 9.1 Aggregation of risks nos. 4 and 11

No.	Incident	Likelihood	Consequence
4	Malware compromises meter data	Rare	Moderate
11	Software bug on the metering terminal compromises meter data	Unlikely	Moderate
4+11	Software on the metering node compromises meter data	Possible	Moderate

shows the results. All original malicious and non-malicious risks are included, as well as risks aggregated from both kinds.

		Likelihood				
		Rare	Unlikely	Possible	Likely	Certain
Consequence	Critical		2			
	Major	6,13	6+13			
	Moderate	4,5	8,11,12,14	4+11,5+12	1	
	Minor			15	3	9
	Insignificant	7			16,17	10

Fig. 9.5 Risk matrix after aggregation

9.4 Risk Grouping

Overviews like the one provided by Fig. 9.5 give an indication of which risks need treatment. However, as preparation for the risk treatment, we also want to take into consideration the fact that treatments may have an effect on several risks, thereby justifying higher cost than if we only consider individual risks. It can therefore be useful to group risks with this is in mind.

The distinction between malicious and non-malicious risks earlier in the assessment has given us two groups. This is already useful, as some treatments will only have an effect on one of these groups. For example, data encryption, firewalls, and intrusion detection systems will usually reduce the likelihood or consequence of (some) malicious risks, without having any effect on non-malicious risks.

In addition to distinguishing between malicious and non-malicious risks, we may typically group risks according to shared vulnerabilities, threats, threat sources, or assets. The purpose of the grouping is to facilitate identification of the treatments that give the best effect for the least cost by placing together risks that may benefit from a common treatment.

In order to find out how to further group risks for our assessment, we systematically go through the results of the risk identification in Sect. 7.2 and Sect. 7.3. Do any of these risks have anything in common that indicates that they will benefit from the same treatment? Here we find, for example, that risk no. 14, *Mistakes during maintenance of the central system disrupt transmission of control data to the choke component*, and risk no. 15, *Mistakes during maintenance of the central system prevent reception of data from metering nodes*, are both related to the threat *Mistakes during update/maintenance of the central system* and to the vulnerability *Poor training and heavy workload*, as illustrated in Table 9.2. As shown in Fig. 9.2,

Table 9.2 Grouping of risks nos. 14 and 15

No.	Incident	Asset	Threat	Vulnerability
14	Mistakes during maintenance of the central system disrupt transmission of control data to the choke component	Provisioning of power to electricity customers	Mistakes during update/maintenance of the central system	Poor training and heavy workload
15	Mistakes during maintenance of the central system prevent reception of data from metering nodes	Availability of meter data	Same as the row above	Same as the row above

risks nos. 14 and 15 are both *Low*, but increasing the likelihood or consequence of either of them by a single step would bring its risk level to *Medium*. Treatments that address both these risks are therefore quite likely to be worth the cost. By grouping such risks we make it easier to take such considerations into account.

Similarly to the above case, we find that risks nos. 4-6 share a common threat and vulnerability, and that the same applies to risks nos. 11-13. Even if each of these risks is part of an aggregated risk with risk level *Medium*, thereby ensuring that they receive attention during the risk treatment, it is still useful to group them together for the purpose of cost-benefit analysis. We therefore create two new groups, one consisting of risks nos. 4-6 and one consisting of risks nos. 11-13.

9.5 Further Reading

For how to deal with uncertainty we refer to Chap. 13, which is dedicated to this particular problem. With respect to risk aggregation and grouping, we are not aware of any standards or similar sources that provide detailed guidelines, although the CORAS method [47] offers some support.

Chapter 10
Risk Treatment

The final step of the cyber-risk assessment starts with identification of treatments for selected risks, as explained in Sect. 2.4.5 and Sect. 5.3.6. We then assess the effect of the treatments and consider whether the residual risk is acceptable. If it is, the documentation is finalized and the process terminates, otherwise we need to go back and do another iteration of the treatment identification.

10.1 Risk Treatment Identification

The techniques we use for treatment identification are to a large degree the same as those described for risk identification in Sect. 7.1, in particular when it comes to obtaining information from standards and repositories, as well as people. In the following we demonstrate treatment identification with respect to malicious and non-malicious risks.

10.1.1 Malicious Risks

Ideally, we would of course like to find treatments for all identified risks. However, since we always have limited time and resources, we need to focus on those that are most important. We therefore start by selecting risks based on the results of the risk evaluation. Here we make sure to include:

- all individual risks that are not *Low* according to the risk evaluation criteria, and
- individual risks that are part of an aggregated risk that is not *Low*.

For the aggregated risks, we prefer to list each individual contributing incident rather than giving a common, more abstract description. This is because the more detailed descriptions of the individual risks can provide information that is useful for coming

© The Author(s) 2015
A. Refsdal et al., *Cyber-Risk Management*, SpringerBriefs
in Computer Science, DOI 10.1007/978-3-319-23570-7_10

up with treatments. However, when evaluating the proposed treatments during the risk acceptance later, we will consider the effect on the aggregated risk.

Table 10.1 shows the result of the selection of malicious risks for which to identify treatments. The second column from the right shows whether the risk is part of an aggregated risk. If so, the aggregation is indicated by a plus sign between the individual risk numbers, as in Fig. 9.5. The rightmost column shows whether the risk is part of a group and, if so, which risks are members of the group. The members of a group are separated by a comma.

Table 10.1 Malicious risks selected for treatment identification

No.	Risk level	Incident	Aggr.	Group
1	High	Data from metering nodes cannot be received by the central system due to DDoS attack	No	No
2	High	False control data received by all or most choke components	No	No
3	Medium	False meter data for a limited number of electricity customers received by the central system	No	No
4	Low	Malware compromises meter data	4+11	4,5,6
5	Low	Malware disrupts transmission of meter data	5+12	4,5,6
6	Low	Malware disrupts the choke functionality	6+13	4,5,6

The next step is to identify treatments for the selected risks. Here we make sure to exploit all the information about threat sources, threats, vulnerabilities, and so on that we obtained during the risk identification, as each of these elements may potentially be targeted by treatments. For each risk we therefore create a small table summarizing this information. Table 10.2 and Table 10.3 show the results for risk no. 1 and risk no. 4, respectively. The final row of the table is dedicated to documenting the treatments that we identify. For risk no. 1, the treatment consists of updating the DDoS detection and response mechanism. This could, for example, be achieved by combining anomaly-based and signature-based detection and classification techniques, and allowing malicious packets to be redirected to a controlled part of the network for analysis, rather than being dropped. For risk no. 4, the treatments consist of frequent updating of malware protection on the metering nodes and strengthening the integrity checking of meter data on the central system. While the former reduces the likelihood of meter data being compromised, the latter will increase the chance that compromised data are detected, thereby allowing the central system operator to take appropriate measures.

We make similar tables for the remaining risks from Table 10.1. These tables are not shown here.

Table 10.2 Treatment identification table for risk no. 1

Element	Description
Risk no.	1
Incident	Data from metering nodes cannot be received by the central system due to DDoS attack
Asset	Availability of meter data
Threat source	Script kiddie; Cyber-terrorist
Threat	DDoS attack on the central system
Attack point	Internet connection to the central system
Vulnerability	Inadequate attack detection and response on central system
Treatment	Implement state-of-the-art DDoS attack detection and response mechanism on central system

Table 10.3 Treatment identification table for risk no. 4

Element	Description
Risk no.	4
Incident	Malware compromises meter data
Asset	Integrity of meter data
Threat source	Malware
Threat	Metering node infected by malware
Attack point	Internet connection to the metering terminal
Vulnerability	Outdated antivirus protection on metering node
Treatment	Frequent updates of malware protection on metering node; Stronger integrity checking of received meter data on central system

10.1.2 Non-malicious Risks

For non-malicious risks we select risks in the same way as we did for malicious risks. Table 10.4 shows the result. Risk no. 9 and risk no. 10 are included due to their individual risk level, while the rest of the risks are included either because they are part of an aggregated risk or a member of one of the risk groups identified during the risk evaluation. Notice that we have decided to include risks nos. 14 and 15, which were grouped together during the risk evaluation, even though each of these risks are *Low*.

For each selected risk we compile the information obtained during the risk identification in a single table to facilitate treatment identification, in a similar way as we did for malicious risks. Table 10.5 shows the result for risk no. 9. Two potential treatments are identified, both of a purely technical nature. The first is to ensure that all electricity customers have a redundant GPRS communication link that can be used in case the Internet connection goes down. The second is to ensure that the choke component does not shut off all power to the electricity customer in the absence of control data. Instead, the default mode should be to allow at least 50% of normal power consumption.

Table 10.4 Non-malicious risks selected for treatment identification

No.	Risk level	Incident	Aggr.	Group
9	High	Communication between the central system and the metering terminal is lost	No	No
10	Medium	Same as the row above	No	No
11	Low	Software bug on the metering terminal compromises meter data	4+11	11,12,13
12	Low	Software bug on the metering terminal disrupts transmission of meter data	5+12	11,12,13
13	Low	Software bug on the metering terminal disrupts the choke functionality	6+13	11,12,13
14	Low	Mistakes during maintenance of the central system disrupt transmission of control data to the choke component	No	14,15
15	Low	Mistakes during maintenance of the central system prevent reception of data from metering nodes	No	14,15

Table 10.5 Treatment identification table for risk no. 9

Element	Description
Risk no.	9
Incident	Communication between the central system and the metering terminal is lost
Asset	Provisioning of power to electricity customers
Threat source	Internet connection to the metering terminal
Threat	Internet connection to the metering terminal goes down
Entry point	Internet connection to the metering terminal
Vulnerability	Single communication channel between central system and metering terminal
Treatment	Install redundant GPRS communication for all electricity customers; Ensure suitable default mode for choke component when communication is lost

For risks nos. 14 and 15, which were grouped together during the risk evaluation, we create a joint table, as shown in Table 10.6. The treatments identified here are of both a human/organizational and technical nature. One option is to simply hire more staff, as heavy workload is recognized as a vulnerability. Another option is to develop executable scripts for performing routine maintenance tasks, which may reduce the likelihood of mistakes during such tasks. Notice that this treatment option could potentially also introduce new risks which must be taken into consideration, for example due to bugs in the scripts. Finally, the last treatment option is to enforce a policy to ensure that only senior personnel are allowed to perform non-routine maintenance tasks.

We create similar tables for the remaining risks from Table 10.4. These tables are not shown here.

Table 10.6 Treatment identification table for risks nos. 14 and 15

Element	Description
Risk no.	14 and 15
Incident	Mistakes during maintenance of the central system disrupt transmission of control data to the choke component; Mistakes during maintenance of the central system prevent reception of data from metering nodes
Asset	Provisioning of power to electricity customers; Availability of meter data
Threat source	Maintenance personnel
Threat	Mistakes during update/maintenance of the central system
Entry point	Central system
Vulnerability	Poor training and heavy workload
Treatment	Hire more staff; Develop executable scripts for routine maintenance tasks; Establish and enforce a policy stating that only senior personnel perform non-routine maintenance tasks

10.2 Risk Acceptance

Implementing treatments always carries a cost, either directly in terms of money or indirectly in terms of, for example, reduced system usability and efficiency, as discussed in Sect. 5.3.6. For each treatment we therefore need to weigh its effect against its cost. We first estimate the effect of a treatment in terms of reduced risk level for the affected risks, before estimating its cost.

Conducting an exact quantitative cost-benefit analysis is not feasible when dealing with the kind of assets and scales that we have defined, and it would be hard to map the consequences to a monetary value. Quantifying the cost of treatments can also sometimes be hard, for example if they involve reduced user-friendliness of systems or security policies affecting the behavior of employees. We therefore decide on a simple, qualitative approach where we adopt the same scale for costs as for risk levels, that is a scale consisting of the steps *High*, *Medium*, and *Low*. The cost-benefit analysis then amounts to comparing costs over this scale with reduction in risk level.

To illustrate the approach, we demonstrate the cost-benefit analysis for some of the treatments identified above. We start with *Implement state-of-the-art DDoS attack detection and response mechanism on central system*. This was identified for risk no. 1, *Data from metering nodes cannot be received by the central system due to DDoS attack*, which has risk level *High* before treatment. Implementing the treatment will hardly prevent script kiddies or cyber-terrorists from launching DDoS attacks, so we do not expect it to have any effect on the threat *DDoS attack on the central system*. However, being able to quickly detect such an attack and respond accordingly will reduce the likelihood that the attack actually leads to the incident in question. Moreover, even if the attack succeeds for a while, a prompt response implies that fewer electricity customers are affected, and that they are affected for a shorter period. We therefore conclude that implementing the treatment will reduce the likelihood of *Data from metering nodes cannot be received by the central system due to DDoS attack* from *Likely* to *Possible* and at the same time reduce its consequence from *Moderate* to *Minor*. This brings the risk level from *High* to *Low*. As risk no. 1 is not part of an aggregated risk or a risk group and the treatment does not apply to any of the other risks, this concludes the analysis of its effect.

To implement the treatment, the central system operator needs to make a significant investment in hardware and network infrastructure to establish a safe and controlled environment where offending packets can be directed for analysis. Moreover, arriving at an adequate set of detectors, preferably combining anomaly-based and signature-based approaches, will take time and effort. The cost of the treatment is therefore *High*. Table 10.7 documents the results of the cost-benefit analysis.

Table 10.7 Effect of treatments

Treatment	Risk	Effect	Cost
Implement state-of-the-art DDoS attack detection and response mechanism on central system	1	High to Low	High
Stronger integrity checking of received meter data on central system	4	Low to Low	High
	11	Low to Low	
	4+11	Medium to Low	
Hire more staff	14,15	Low to Low	High
Develop executable scripts for routine maintenance tasks	14,15	Low to Low	Low

We now move on to risk no. 4, *Malware compromises meter data*, which is part of the aggregated risk 4+11. According to Table 10.3, the treatments *Frequent updates of malware protection on metering node* and *Stronger integrity checking of received meter data on central system* were identified for this risk. Here we illustrate the considerations regarding the latter.

The treatment *Stronger integrity checking of received meter data on central system* is expected to reduce the consequence of the incident. The reason is that recognizing false meter data early allows the central system operator to discard these data and implement corrective measures. Power consumption can to a large degree be predicted from historical data to which the central system operator has access. As soon as received meter data are recognized as false, it is therefore possible to obtain a good approximation of the correct data, which can adequately serve the needs of the central system operator until the situation is restored. We therefore estimate that implementing the treatment reduces the consequence of risk no. 4 from *Moderate* to *Insignificant*, while the likelihood is unchanged. This will, of course, not reduce the risk level of risk no. 4, which is already *Low* before any treatments.

However, we also need to consider whether the treatment affects the risk level of the aggregated risk 4+11, which has risk level *Medium* before treatment. Here we notice that *Stronger integrity checking of received meter data on central system* is equally good for detecting meter data compromised by software bugs and for detecting data compromised by malware. We therefore estimate that the consequence of risk no. 11 is also reduced from *Moderate* to *Insignificant*. Consequently, we also set the consequence of the aggregated risk 4+11 to *Insignificant*, which takes its risk level from *Medium* to *Low*.

It remains to estimate the cost of the treatment. On the central system side the cost is fairly low. Unfortunately, for certain types of metering terminals, implementation of the treatment requires upgrading of the hardware. The cost is therefore set to *High*.

In addition to the above, in Table 10.7 we have also included two of the treatments for the group consisting of risk no. 14 and risk no. 15, even if each of these risks has risk level *Low* before treatment.

Notice that different treatments may affect each other, either by reinforcing each other or to some extent canceling each other out. We need to take this into account in the cost-benefit analysis. In such cases we can add separate entries for the potential treatment combinations and estimate each combination as if it was an individual treatment. Moreover, with respect to costs of treatments we make sure to take maintenance into account if this is relevant. We also need to consider whether treatments may introduce new risks, as discussed earlier concerning the introduction of executable scripts to address risks nos. 14 and 15.

After performing the cost-benefit analysis, it remains to decide which treatments to implement and whether the residual risk is acceptable. In the end, these decisions must be made by the decision makers of the organization for which the assessment is performed. We terminate the process by recording the decisions and finalizing the documentation.

10.3 Further Reading

ISO/IEC 27032 [28] comes with a list of cybersecurity controls that can be utilized for treatment identification. The data breach investigation report by Verizon [82] also provides an overview of critical security controls mapped to incident patterns which can support the identification of treatments. The OWASP overview of the ten most critical web application security risks [63] offers advice on prevention.

The CORAS method [47] provides further advice on the kind of cost-benefit analysis adopted in this section. Moreover, there is a CORAS extension [4] offering techniques and guidelines to establish compliance with ISO/IEC 27001.

Part III
Known Challenges and How to Address Them in Practice

Chapter 11
Which Measure of Risk Level to Use?

So far in this book we have measured the risk level of incidents in terms of consequence for assets and likelihood of occurrence. In other words, we have measured risk level based on two factors, namely loss of asset value when a potential incident occurs and how often this happens. In this chapter we present and discuss alternative ways of measuring risk level using two, three, or even more factors.

11.1 Two-factor Measure

A risk is often expressed in terms of a combination of the consequences of an event and the associated likelihood of occurrence, where consequence is the outcome of an event affecting assets. This is the classical two-factor measure of risk.

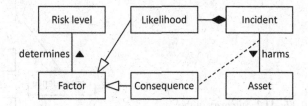

Fig. 11.1 Summary of two-factor approach

Figure 11.1 illustrates the two-factor approach using the UML [58] class diagram notation. Each line connecting two boxes represents a relation. The white-headed arrows pointing from likelihood and consequence to factor imply that the concepts likelihood and consequence should be understood as instances of the more general concept factor. In other words, they are both factors. Moreover, the factors determine the risk level.

© The Author(s) 2015

A. Refsdal et al., *Cyber-Risk Management*, SpringerBriefs
in Computer Science, DOI 10.1007/978-3-319-23570-7_11

The relation with a black diamond connecting incident and likelihood captures that likelihood is an attribute of incident. This is because likelihood is a measure of incident occurrence. On the other hand, consequence is connected to the relation between incident and asset since it is a measure of the former's potential to affect the latter.

11.2 Three-factor Measure

In the field of security, three-factor risk measures are popular. For example, NIA-CAP [57] defines risk as "a combination of the likelihood that a threat will occur, the likelihood that a threat occurrence will result in an adverse impact, and the severity of the resulting impact."

The "likelihood that a threat will occur" is a measure of the extent to which the target is subject to a certain threat, while the "likelihood that a threat occurrence will result in an adverse impact" is a measure of the vulnerability of the target with respect to the threat in question. These two factors may be understood as a decomposition of likelihood from the two-factor approach since the likelihood of a threat occurring and the likelihood of it resulting in an adverse impact may be used to deduce the likelihood of a risk in the two-factor sense. The third factor "severity of resulting impact" corresponds to consequence in the two-factor case. The meaning of "combination" is not further defined. Hence, we may think of the risk level as a triple of factors whose relative weighting is left open.

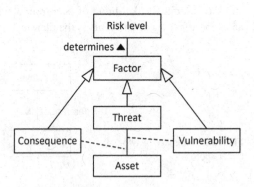

Fig. 11.2 Summary of three-factor approach

The definition is summarized in Figure 11.2, again using a UML class diagram. We now distinguish between three factors, namely threat, vulnerability, and consequence. As in the case of consequence, vulnerability is connected to the relation between threat and asset since it is a measure of the threat's potential to harm the asset.

11.3 Many-factor Measure

In some situations it may be beneficial to use even more factors. OWASP [64], for example, which is concerned with the security of web applications, recommends an approach where the likelihood of the two-factor approach is decomposed into threat agent factors and vulnerability factors. Similarly, consequence is represented by technical impact factors and business impact factors. The proposed vulnerability factors with respect to a group of attackers are, for example:

- Ease of discovery: How easy is it for this group of attackers to discover this vulnerability? Practically impossible (1), difficult (3), easy (7), automated tools available (9).
- Ease of exploit: How easy is it for this group of attackers to actually exploit this vulnerability? Theoretical (1), difficult (3), easy (7), automated tools available (9).
- Awareness: How well known is this vulnerability to this group of attackers? Unknown (1), hidden (4), obvious (6), public knowledge (9).
- Intrusion detection: How likely is an exploit to be detected? Active detection in application (1), logged and reviewed (3), logged without review (8), not logged (9).

According to OWASP, in the case of threat agent and vulnerability factors, the numbering from 0 to 9 is a likelihood rating. The overall likelihood is formally defined as the average of the likelihood factors. Similarly, the overall consequence is equal to the average of the technical impact and business impact factors. The risk level is then defined via a risk matrix as in the two-factor case.

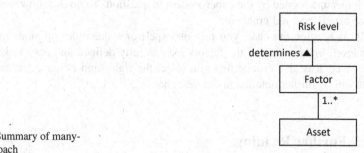

Fig. 11.3 Summary of many-factor approach

Figure 11.3 illustrates a many-factor measure, again as a UML class diagram. The factors, of which there may be any finite number, are all of relevance for the asset, and for representing and measuring risk level with respect to the asset.

11.4 Which Measure to Use for Cyber-risk?

As we have seen, risk level may be measured in multiple ways. The same holds
for cyber-risk. We have presented the two-factor approach based on consequence
and likelihood. The two-factor approach is the one most commonly used in practice, also within cybersecurity. We have also considered one of several alternative
approaches employing three factors developed for the security domain. Finally, we
have discussed the use of more than three factors.

Which approach you should use and how you should use it depends on the context and your risk assessment situation. What data is available is an important parameter when deciding how to measure risk level. If you have good data on frequency
and consequence you will probably go for the two-factor approach, and accordingly
for other measures if they are favored by the data available.

Within cybersecurity our impression is that the popularity of approaches using
more than two factors is growing. One reason is that measuring likelihood with a
reasonable degree of uncertainty in practice may be difficult. Consider, for example,
an attack by a malicious threat source on some given target. It may be the case
that the likelihood of a successful attack depends almost entirely on the motive and
abilities of the attacker, in addition to the vulnerabilities of the target with respect
to the attack in question. If these factors are easy to measure within acceptable
uncertainty, you may use them directly to calculate the risk level, instead of going
indirectly via likelihood.

Most cyber-systems generate logs automatically with respect to a (large) number
of indicators. Hence, when assessing risk, the problem normally is not the lack of
data, but the lack of the right kind of data with respect to predefined factors. In
such situations you may try to define your own risk function from factors matching
the indicators logged by the cyber-system in question. To do this, however, requires
some experience and great care.

If, as is often the case, you rely on expert or stakeholder opinions to estimate
risk level, make sure that the factors are carefully defined and easy to keep apart.
Moreover, it is also crucial that you select the right kind of scale for each factor.
This will be further detailed in the next chapter.

11.5 Further Reading

Section 11.2 employs NIACAP [57] to exemplify a three-factor risk measure. Three-
factor measures are not specific to cybersecurity. Risk of terrorism, or malicious attack in general, is often measured accordingly [85]. In Sect. 11.3 we use the OWASP
approach [64] as an example of a many-factor risk measure. The number of factors
and how the factors are decomposed vary. While OWASP describes the attacker in
terms of skill level, motive, opportunity, and size, the Common Criteria [8] employs the elapsed time (for the attack), expertise, knowledge of target, window of
opportunity, and equipment (for the attack).

Chapter 12
What Scales Are Best Suited Under What Conditions?

A core aspect of risk assessment is to prioritize risks according to their risk level. This requires a measure of risk level. As explained in Chap. 11, the risk level depends on (or is a function of) several factors. Independent of which and how many factors, we need a suitable scale for each of them. The suitability of a scale depends of course on the factor in question, but also on the kind of risk assessment we are conducting, the target of assessment, and the available sources of data. In this chapter we first give a classification of scales. Then we explain the difference between qualitative and quantitative risk assessment, but also why qualitative and quantitative risk assessment may meaningfully be combined. Thereafter we address the suitability of different scales for measuring likelihood and consequence, respectively. Finally, we consider scales from the perspective of the specific challenges of cyber-risk.

12.1 Classification of Scales

Building on [74], we distinguish between four basic categories of scales:

- Nominal scales – allow the determination of equality.
- Ordinal scales – allow the determination of greater or less.
- Difference scales – allow the determination of equality of intervals or differences[1].
- Ratio scales – allow the determination of equality of ratios.

A *nominal scale* allows us to categorize the phenomena we are concerned with into disjoint categories. We may, for example, classify threat sources according to

---

[1] Referred to as interval scales in [74] and also in much other literature. We prefer the term "difference scales" to avoid difference scales being confused with scales of intervals. We may define a scale of intervals on the basis of an ordinal scale. Consider for example, the ordinal scale consisting of the seven values a, b, c, d, e, f, g and assume they are ordered alphabetically. The scale consisting of the three intervals $[a,b], [c,e], [f,g]$ is a scale of intervals, but not a difference scale.

© The Author(s) 2015
A. Refsdal et al., *Cyber-Risk Management*, SpringerBriefs
in Computer Science, DOI 10.1007/978-3-319-23570-7_12

the nominal scale consisting of the following three categories: "non-human threat source," "accidental human threat source," and "intentional human threat source." An *ordinal scale* is a nominal scale whose categories or values are ordered. We may use an ordinal scale to order the phenomena we are measuring. We may for example measure the impact of a risk with respect to whether the impact is "minor," "moderate," or "major." A *difference scale* is an ordinal scale such that equality of difference at the level of values implies that the corresponding phenomena are equally distinct. The Celsius scale is a difference scale. A $10\,^{\circ}$C difference between two correct measurements $m\,^{\circ}$C and $m'\,^{\circ}$C reflects the same temperature difference in reality independent of whether $(m, m') = (-10, 0)$ or $(m, m') = (100, 110)$.

A *ratio scale* is a difference scale such that equality of ratios at the level of values implies that the corresponding phenomena are equally distinct. An example of a ratio scale is "the number of identity thefts in Europe per year." Note that the Celsius scale is not a ratio scale because it does not make sense to say that $10\,^{\circ}$C is twice as warm as $5\,^{\circ}$C since $0\,^{\circ}$C is "arbitrarily" selected as the freezing point of water under certain conditions. The Kelvin scale on the other hand is a ratio scale. A special case of a ratio scale is an absolute scale [18]. An *absolute scale* is a ratio scale in which one value among the values may be understood as the smallest unit. For example, 1 is the smallest unit among the natural numbers.

A qualitative risk assessment makes use of nominal and ordinal scales, and the values are typically captured by natural language expressions. In fact, most of the scales will be ordinal for the simple reason that risk evaluation requires a form of ordering. After all, a risk evaluation is supposed to end up with a risk prioritization. In quantitative risk assessment the values are numbers and the scales are mostly of the ratio kind. Moreover, a quantitative risk assessment may also use difference scales or absolute scales. In the following section we address the relationship between qualitative and quantitative risk assessment in further detail. Hence, we refer to ratio and difference scales as *quantitative scales*, while scales that are (only) nominal or ordinal are called *qualitative scales*.

12.2 Qualitative Versus Quantitative Risk Assessment

We start by exemplifying a qualitative approach. Using an ordinal scale, the consequence of a confidentiality breach for an aircraft company may be measured according to the consequence scale given in Table 12.1. Each consequence value in the table is described by an expression in English. This scale using qualitative values is not easily converted into an equally useful ratio scale. The reason is that we would have to quantify information that is not homogeneous. In general, a quantitative approach tends to work better when conducting the assessment at a more technical level or with a fine level of granularity. If, for example, the asset in question is a homogeneous customer database, then we may measure the impact of information leakage by "the number of customer records being leaked." This would be a quan-

Table 12.1 Consequence scale for confidentiality using qualitative values

Consequence	Description
Catastrophic	Leakage of data that can be utilized in terror
Major	Data leakage with legal implications
Moderate	Distortion of aircraft company competition
Minor	Leakage of aircraft information data
Insignificant	Leakage of publicly available data

titative scale consisting of the natural numbers less than or equal to the number of records in the database.

We may also use an ordinal scale to measure likelihood, as exemplified by Table 12.2. Each value is characterized by language expressions. Notice that we may

Table 12.2 Likelihood scale using qualitative values

Likelihood	Description
Certain	A very high number of similar occurrences already on record; has occurred a very high number of times at the same location
Likely	A significant number of similar occurrences already on record; has occurred a significant number of times at the same location
Possible	Several similar occurrences on record; has occurred more than once at the same location
Unlikely	Only very few similar incidents on record when considering a large traffic volume or no records on a small traffic volume
Rare	Has never occurred yet throughout the lifetime of the system

explain or exemplify values using two or more statements if desired. A corresponding quantitative scale could be one based on frequencies; for example, "the number of occurrences per year." Using the conventional two-factor approach, the risk function is a mapping from a pair of consequence and likelihood values to risk level. Within qualitative risk assessment the risk function is normally defined by the risk matrix. The risk levels may for example be captured according to the ordinal scale in Table 12.3.

Table 12.3 Scale of risk level using qualitative values

Risk level	Description
High risk	Unacceptable and must be treated
Medium risk	Must be evaluated for possible treatment
Low risk	Must be monitored

Within quantitative risk assessment the risk level is commonly described as the product of the risk factors. With respect to the quantitative scales proposed above, the risk level would be "the number of records leaked" multiplied by "the number of occurrences per year." In practice the difference between the qualitative and quantitative schools of risk assessment is not as fundamental as the respective proponents like to argue. In fact a combination of the two is commonly what is required. A purely qualitative approach is not satisfactory on its own because the result of any risk assessment must be converted to a quantitative scale in some way or another in the end. After all, whether a risk is acceptable or not is in the end a matter of cost measured in monetary value. A purely quantitative approach, on the other hand, is often "overkill." Working with exact quantities is fine in theory but not necessarily very practical. For example, the quantitative scale for risk level based on multiplication may yield infinitely many values of risk level, although perhaps three or five would be sufficient, since the number of decision alternatives is typically small. If, for example, the purpose of a risk evaluation is to decide whether a risk has a sufficiently low risk level to be left untreated or should be considered further in a cost-benefit analysis, strictly speaking we need only two risk values.

12.3 Scales for Likelihood

Estimating or measuring likelihood tends to be difficult. As we discuss more generally in Chap. 13, one reason is that there may be considerable uncertainty as to what the likelihood is. Another reason, as we address in Chap. 14, may be the lack of experience or historical data with respect to the event in question. For example, the event may not have occurred in the lifetime of the target of assessment, or perhaps not yet at all. A third reason, which we focus on in the following, is that we may all too easily complicate the task ourselves by selecting quantitative likelihood scales that are badly suited to the task.

In general, we do not recommend using probabilities when interacting with people in a risk assessment situation. A probability is always defined implicitly with respect to some time interval or context, and the existence of this implicit interval or context is easily overlooked or misunderstood, leading to bad estimates and confusion. Our experience is that intervals of frequencies or qualitative scales work best in practice. In fact, there is considerable empirical evidence all the way back to the 1960s showing that human beings, experts as well as laymen, easily get confused when being exposed to statistical arguments based on probabilities [43]. This is a problem when we as risk assessors communicate our results, but also when we try to obtain data from human beings in terms of probabilities or conditional probabilities. Numerous experiments [20] have shown that human beings comprehend natural frequencies better than probabilities. Natural frequencies are the result of an information-sampling process consisting of observing and counting events. Said differently, natural frequencies are the result of the sequential partitioning of one total sample into subsamples. Figure 12.1 illustrates the idea. Assume we have a sample of 622 different incidents. 319 of these 622 were caused by external threats

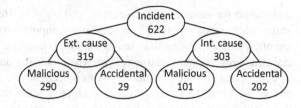

Fig. 12.1 Natural frequencies exemplified

while 303 had internal origin. The incidents may be further classified according to whether they were caused by intent or by accident.

12.4 Scales for Consequence

A risk evaluation may end up with a cost-benefit analysis or be used as input for cost-benefit analysis. To simplify, for example, the calculation of the annualized loss expectancy, we might prefer or be under pressure to measure consequence in terms of monetary value. Depending on the kinds of assets to be protected, this may, however, not work well in practice. There are of course some assets that are well-suited to monetary scales. If, for example, the asset in question is a bag of diamonds the consequence of an incident in which some or all of the diamonds are stolen might be equal to the monetary value of the diamonds stolen. On the other hand, if the asset is the integrity of a customer database, it may be easy to characterize the number of records harmed, but hard to say what this means in euros.

The situation becomes even worse if the asset in question is something as volatile as a company's reputation. In most cases it is hard to know or characterize the impact of some incident on reputation, and even harder to estimate what this impact corresponds to in euros. A company's reputation is influenced by many different factors, and how they add up or counter each other is hard to predict. The suitability of a consequence scale obviously depends on the asset in question. In fact, we recommend defining specialized consequence scales for each asset of relevance. Furthermore, a consequence scale has to be defined in such a way that it fits its intended usage. A consequence scale suited for communicating consequences to decision makers may be unsuited to discussions with technical people. Hence, an experienced risk assessor may measure consequences in different ways for the same asset depending on who is being approached.

12.5 What Scales to Use for Cyber-risk?

If we restrict our attention to cyber-risk, some scales are simpler to define while others are more challenging. The main simplifying feature is that cyber-risk concerns systems which to a large extent are computerized. The computerized parts are well-suited for automatic measurement and logging. Hence, to the extent that the scales address computational features at a computational level of abstraction they are often easier to define than scales aiming to measure aspects of for example social structures. A complicating feature when focusing on cyber-risk is the openness of cyberspace and the fact that it is often necessary to measure human intentions and skills. Table 12.4 exemplifies a qualitative scale that can be used to measure the skill level of an attacker. It is not easy to measure skill level quantitatively. Of

Table 12.4 Scale of attacker skill level using qualitative values

Skill level	Description
Specialist	Security penetration skill
Advanced	Network and programming skill
Good	Experienced computer user
Some	Some technical skill
None	No technical skill

course, you may use the numbers 1-5 instead of the lexical values. However, you would still depend on the descriptions to explain what the numbers denote. Hence, the scale would still be ordinal. Also nominal scales may be of great value. You may for example find it difficult to order the intentions or motives of threat sources. To classify them using a nominal scale as defined by Table 12.5 may be helpful when, for example, identifying and estimating the likelihood of threats and incidents.

Table 12.5 Scale of motive of threat sources

Motive	Description
Profit	Earn money
Challenge	Obtain satisfaction because it is difficult
Protest	Raise some political issue
Enjoyment	For the fun of it
Revenge	Pay back some injustice

12.6 Further Reading

Scales and measurements are topics of relevance for most human activities and there is a huge literature available spread over many fields. Johan Galtung's book [18], for example, focuses on social research, while ISO/IEC 25010 [31] on software quality addresses the domain of software engineering. With respect to cybersecurity there are many proposals, see for example, ISO/IEC 27004 [29] on security measurement, as well as OWASP [64] and the Common Criteria [8]. Moreover, the field is so far not very consolidated.

Chapter 13
How to Deal with Uncertainty?

ISO 31000 [25] defines uncertainty to be "the state, even partial, of deficiency of information related to, understanding or knowledge of an event, its consequence, or likelihood". If you find this definition hard to comprehend, we are not surprised because so do we. A meaningful interpretation is that we may be uncertain about the extent to which a measurement or estimate of consequence is correct; we may also be uncertain with respect to a likelihood, or the measurement of any other factor used to capture the risk level. In fact, there may also be uncertainty related to the exact nature of the risk itself: What kind of phenomenon are we dealing with? In what way does it differ from other phenomena? How and in what way can it materialize?

This chapter distinguishes between two kinds of uncertainty, namely epistemic and aleatory uncertainty. Furthermore, it relates these concepts to the terminology introduced in previous chapters. Thereafter, we focus on how to represent uncertainty and how to reduce it. Finally, we address challenges related to uncertainty in the setting of cyber-risk assessment.

13.1 Conceptual Clarification

As illustrated by Fig. 13.1, uncertainty may concern an incident; for example, its real nature or exact properties. Uncertainty may also concern the factors that are used to measure risk, which may be two as in the two-factor approach, three as in a three-factor approach, or more. The black arrowheads specify the reading direction

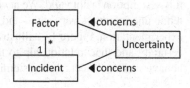

Fig. 13.1 Uncertainty in relation to the risk terminology

© The Author(s) 2015
A. Refsdal et al., *Cyber-Risk Management*, SpringerBriefs
in Computer Science, DOI 10.1007/978-3-319-23570-7_13

of a relation. This means that uncertainty concerns factor. Hence, if we measure risk in terms of likelihood and consequence we may have uncertainty with respect to both. Moreover, the uncertainty with respect to the two factors may be different. It may be that we are almost certain that the consequence measurement is correct, but uncertain with respect to the correctness of the likelihood estimate, or the other way around. Similarly, if we use a three-factor approach we may have different uncertainties with respect to the measurement of the different factors; for example, with respect to threat, vulnerability, and consequence, as illustrated by Fig. 11.2.

13.2 Kinds of Uncertainty

Some literature, see for example Tony O'Hagen's paper [59], distinguishes between two kinds of uncertainty. On the one hand we may be uncertain due to ignorance or lack of evidence. This kind of uncertainty is commonly referred to as *epistemic uncertainty* and pertains to our knowledge about the object at hand. On the other hand uncertainty may be due to inherent randomness. The latter kind of uncertainty is commonly referred to as *aleatory uncertainty* and pertains to chance. Typical examples are the outcomes of the tossing of a coin, or the hands players of a game of poker receive. Aleatory uncertainty is the inherent randomness that cannot be removed without redesigning the object at hand.

Since the literature operates with different notions of uncertainty, we have to carefully explain how uncertainty should be understood in this book. In fact, what ISO 31000 refers to as uncertainty may best be understood as epistemic uncertainty while aleatory uncertainty may be determined from what ISO 31000 refers to as likelihood. In determining future behavior, we may identify the possible outcomes of a situation and assign a probability $p \in [0, 1]$ to each outcome. In cases of perfect knowledge and where the likelihood p is close to 0 or 1, the outcome is almost certain; there is no epistemic uncertainty and close to no aleatory uncertainty. For example, in a lottery with a million tickets and only one winner, each individual ticket is almost certain to loose. However, if the likelihood p gets closer to 0.5, the outcome is increasingly uncertain, as for example the outcome of tossing a coin. In this case the aleatory uncertainty is high.

In the case of perfect knowledge there is no epistemic uncertainty. This means that there is no further knowledge to be gained from additional empirical investigations. On the other hand, in the case of imperfect knowledge there is always some epistemic uncertainty. For example, in the case of a coin toss, the likelihood is 0.5 only under the assumption that the coin is 100 % perfect and symmetric. In practice, this assumption is not valid. We nevertheless continue to toss coins because the epistemic uncertainty is very small and almost impossible to determine without careful experimentation. Figure 13.2 illustrates the distinction between the two kinds of uncertainty and their relationship to the ISO 31000 terminology. When studying the behavior of an object, the epistemic uncertainty is something we actively seek to

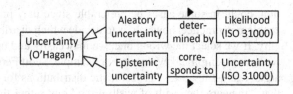

Fig. 13.2 Two kinds of uncertainty

reduce by gathering more information or evidence. In the rest of this book, we use the term uncertainty to mean epistemic uncertainty.

13.3 Representing Uncertainty

Using quantitative scales, we may from a practical point of view represent uncertainty by the use of intervals. The intuition is then that the width of the interval specifies the level of uncertainty. In the case of a coin toss, we know that the likelihood is close to 0.5 but not exactly how close. However, we are perhaps pretty sure that the deviation is not more that 2 percentage points. This means that the likelihood lies somewhere within the interval $[0.48, 0.52]$. The wider the interval, the more uncertainty. When new knowledge is gathered, the uncertainty may be reduced. The result is a narrower interval and a better prediction.

In a risk assessment, whether the level of uncertainty is tolerable often corresponds to the extent to which the uncertainty impacts the decision procedure. Consider for example the risk matrix in Fig. 13.3. Since the levels of uncertainty are represented by the width of the intervals, the larger the box the more uncertainty. The uncertainty related to the risks $r1$ and $r5$ is unproblematic since it does not impact the risk value. This follows since their boxes are positioned within a uniformly shaded area. The same holds for $r2$. Which shading $r3$ belongs to depends on the consequence argument. Hence, in the case of $r3$ only the uncertainty of the consequence argument impacts the risk value, while the risk value of $r4$ depends on the uncertainty related to both arguments. Intervals provide no information except upper

		Likelihood				
		Rare	Unlikely	Possible	Likely	Certain
Consequence	Critical				r2	
	Major			r3		
	Moderate	r1	r5		r4	
	Minor					
	Insignificant					

Fig. 13.3 Uncertainty in risk matrix

and lower bounds. Theoretically, using distributions, such as the Gaussian distribution, could in some cases be preferable since they provide more information. The problem in practice, however, is to decide which distribution better approximates reality. If we select the wrong one, we may easily end up with conclusions for which we have no empirical evidence. For example, Scott Ferson and Lev R. Ginzburg [17] argue against the use of probabilistic distributions for exactly this reason, claiming that: "they are the result of wishful thinking, rather than a careful analysis of what is actually known. This example illustrates what may be a widespread problem with applying classical probability theory in risk assessment where the relevant empirical information is sorely incomplete (as is usually the case)." Moreover, we must also consider whether the additional information deduced, if correct, is of any help or relevance for the decision maker.

Using qualitative scales, we recommend representing uncertainty separately; for example, as a separate natural language expression for each measurement according to an ordinal scale. We illustrate this is in Table 13.1 which documents the consequence of incidents as well as our uncertainty regarding their correctness. Without

Table 13.1 Consequence of incidents with uncertainty estimates

Incident	Consequence	Uncertainty
Information leakage	Low	Some
Breakdown of server	High	No
Identity theft	High	Considerable
Spyware installed	Medium	Some

the uncertainty estimates we might arrive at the conclusion that *Identity theft* is very serious. However, with the uncertainty estimate *Considerable* this does not follow straightforwardly anymore. On the other hand, since there is no uncertainty regarding *Breakdown of server* we know that this incident has *High* consequence.

13.4 Reducing Uncertainty

In cases where the level of uncertainty impacts the decision process, we may try to reduce it. To reduce uncertainty based on the data we have already acquired we may use alternative assessment methods. There exists a plethora of different methods to analyze and reason about uncertainty. Fuzzy logic, for example, has been suggested as a means to analyze vagueness due to uncertainty and possibly deduce stronger, but nevertheless valid conclusions [22]. If this is not sufficient we may attempt to obtain more data. A well-known strategy to reduce uncertainty is to carry out repetitions. If, for example, our risk model is compiled on the basis of interviewing n stakeholders individually, we may conduct interviews of m additional stakeholders and hope thereby to reduce the uncertainty. Repetition is of course not the only possibility. We might, for example, gather the n stakeholders together in a meeting room and ask them to discuss their deviations, and perhaps thereby obtain more accurate estimates. We might also try to reduce the uncertainty by testing the risk model against historical data or by conducting various surveys across larger groups of stakeholders. This way of trying to measure the same thing using different methods is often referred to as triangulation.

13.5 How to Handle Uncertainty for Cyber-risk?

What we have said regarding uncertainty so far is also valid for cyber-risks. Since cyber-systems to a large extent are computerized, there are many potential data sources and possibilities for triangulation. The monitoring infrastructure of computerized systems may be used to gain additional knowledge regarding likelihoods and how scenarios may develop. Furthermore, if the uncertainty estimate depends on the hypothesized presence of specific software vulnerabilities, we may test the software with respect to the existence of these vulnerabilities.

The uncertainty is often due to lack of knowledge as to whether the vulnerabilities in question actually exist, and if so, how easily they can be exploited. In order to gain more information we may test. A risk model describing the scenarios leading to risks for which the uncertainty is too high, is well-suited to be a starting point for testing. If we find the vulnerabilities exploited in the scenarios, we know for sure. If not, they may still exist, but we may have good reasons to reduce the uncertainty with respect to the likelihood (as well as the likelihood itself).

A considerable source of uncertainty is the dependency upon a cyberspace whose evolution is not very predictable. There are, however, government bodies (such as NIST), multinational institutions (such as ENISA), and independent organizations (such as OWASP) continuously reporting on the security relevant aspects of cyberspace. A good risk assessor will make use of these for all they are worth and in addition inspect logs from the attack surface to look for new patterns and trends.

13.6 Further Reading

Uncertainty is related to topics like subjectivity, trust, and accuracy of measurement [72]. We have already mentioned fuzzy logic [2]. In fact there is a wide variety of approaches and techniques to handle uncertainty, such as Bayesian belief networks [40], Dempster-Shafer structures [71], and subjective logic [41].

For a systematic literature review on the combined use of risk assessment and testing we refer to [10]. Experience from using security testing to validate security risk models in two industrial risk assessments is presented in [11].

Chapter 14
High-consequence Risk with Low Likelihood

High-consequence risk with low likelihood is a challenge within risk management in general. And even more so within the domains of risk management, like cyber-security, where human intentions and behavior are important. This challenge is, however, not just one challenge, but rather a family of related challenges that should be treated separately. We distinguish between:

- Incidents that occur as complete surprises without ever having been considered; for example, something almost unthinkable that has never happened before, like al Qaeda's attack in the USA on September 11, 2001.
- Incidents whose unlikely occurrences are just likely enough to allow them to be anticipated; for example, the awakening of a volcano whose last eruption occurred 500 years ago.

In the literature, the former are commonly referred to as "black swans," while "gray swans" is used to denote the latter.

In our opinion risk assessment is of little help in identifying black swans. Risk assessment is basically a tool for obtaining a consistent picture of risk knowledge already available implicitly or explicitly within the context of the target of assessment. This knowledge is in most cases insufficient to identify black swans, but sufficient in the case of gray swans. In the following we first address black swans and then gray swans.

14.1 Dealing with Black Swans

A *black swan* is an incident that is extremely rare and unexpected, but has very significant consequences [49]. The "black swan" metaphor has historical origins. In sixteenth century Europe it was generally accepted that all swans were white. In fact, "black swan" was used in every day speech as a metaphor for something unthinkable. When sailing up a river in Western Australia January 10 1697, the members of Willem de Vlamingh's expedition were therefore highly surprised when

© The Author(s) 2015

A. Refsdal et al., *Cyber-Risk Management*, SpringerBriefs
in Computer Science, DOI 10.1007/978-3-319-23570-7_14

they observed swans that were black [84], probably the first Europeans to do so. Tellingly they named the river "Swan River." Nevertheless, the black swan metaphor has survived over the centuries, and recently its popularity has increased due to the writings of Nassim Nicholas Taleb [77]. Black swans are also often referred to as "unknown unknowns."

Risk assessors are regularly confronted with questions related to the coverage of their methods, and in particular their ability to discover black swans. In our view, black swans are not likely to be discovered by risk assessment. Risk assessment is basically good for putting together and structuring a consistent picture of explicit and implicit knowledge already residing within the context of a target, not for picking up the unexpected of which there is no knowledge.

This does not mean that there is nothing we can do to prepare for black swans, only that risk assessment is not the right tool for doing it. Black swans will occur and may harm even the most risk averse organization. Hence, independent of how carefully we as risk assessors conduct our risk assessments, we must be humble and communicate to our customer (the party on whose behalf we are assessing) that although the risk picture we deliver is as good as we can make it given the available resources and information, it may be incomplete and they still need to prepare for the unexpected and plan for the unknown. Hence, risk assessment does not make contingency planning, the act of preparing and planning for major incidents and disasters, obsolete. In fact, as we see it, developing good contingency plans is the best approach to cope with black swans.

14.2 Identifying Gray Swans

A *gray swan* is an incident which has far-reaching consequences, but, unlike a black swan, can be anticipated to a certain degree [50]. If we are not careful, gray swans may also easily be overlooked in a risk assessment situation because they are not present in the documentation that we as risk assessors have gained access to. They may also be overlooked because we are not interacting with the right group of stakeholders or because we are not able to extract the required information. Although gray swans may be very unlikely to happen in the short term, they may in principle occur within the next hour, and with grave consequences. Hence, their identification is essential.

By definition, there is knowledge of all relevant gray swans within the context of the target of assessment, if not explicitly written down in some document then at least implicitly within the mind of a stakeholder or deducible from the available data. Our task as risk assessors is to extract this knowledge so that it becomes a part of our risk assessment. Whether we succeed or not depends on our approach to risk assessment and the resources available to us.

In order to uncover the true nature of some phenomenon it is often fruitful to try to observe this phenomenon from different viewpoints. This has several implications for how we conduct risk identification. One implication if we conduct the risk identification based on interviews or workshops, is that we should try to make sure that the group of people we are interacting with contains representatives for each relevant stakeholder role. After all, the perspective of a decision maker, for example, is quite different from that of an internal software developer, which again is different from that of an external consultant hired in for a period of say six months.

Another implication is that given the availability of the necessary resources and budget, we may split the risk identification into a set of independent processes each taking a slightly different approach. In one process we might, for example, start from the assets and try to identify how they may be harmed; in another process the starting point could be known threats and their potential for attacking the target; a third alternative would be to start from known vulnerabilities; and so on. A third implication is that it might be a good idea to embed the use of different kinds of tools in the risk identification process, each providing a new perspective on the target. We may, for example, use a combination of automatic vulnerability scanners, penetration tests, and monitoring tools to provide input to the risk identification.

14.3 Communicating Gray Swans

Consider a gray swan for which we assume there is a very small and exact likelihood, meaning close to no uncertainty, such as the likelihood of an attacker guessing an eight character password by sheer luck. In this case, our challenge as risk assessors boils down to the problem of communicating a very small number in such a way that its size is fully comprehended by those required to act upon it, namely decision makers.

How successful we are at this task may have great impact on the success of the decision finally made. As already explained in Sect. 12.3, we should avoid using probabilities; natural frequencies are more likely to be understood. How we present a frequency also matters a lot. For example, the frequency 0.000005 per year corresponds to the frequency 1 in 200,000 years which is the same as the frequency of once since the beginning of the human race. In general, it makes sense to present very small (and very large) numbers by relating them to entities providing a suitable perspective. On the other hand, if there is considerable uncertainty as to how small the likelihood of the gray swan is, then also this uncertainty must be communicated to relevant decision makers; for example, in terms of an upper and lower bound. In this case, there will be two small numbers to communicate. A good strategy is again to relate the numbers to some entity of quantity providing intuition, and in such a way that also the size of the uncertainty interval is fully comprehended. For example, zero to five times since the beginning of the human race.

14.4 Dealing with Gray Swans

Assume we are in the treatment phase, trying to aid the decision makers in making the right decision regarding a gray swan. If the gray swan is treatable at low cost, then it should clearly be treated. Unfortunately, this is normally not the case. For some gray swans the consequences are so grave that reducing the likelihood is the only viable alternative, even at very high cost. There are however also gray swans for which this is financially unfeasible and then the option left is to try to reduce the consequence. One obvious strategy to avoid cyber-attacks on critical infrastructure is to disconnect the critical infrastructure from cyberspace. However, in most cases this is not financially feasible. A specialized contingency plan may be a good option. A gray swan, in contrast to a black swan, is something we have knowledge or experience about in one form or another. The contingency plan for a gray swan may therefore be much more specialized and also much more effective than a contingency plan for black swans.

14.5 Recognizing Gray Swans in Cyberspace

The challenge of estimating high-consequence risks with very low likelihood is a family of challenges. What is said above regarding black and gray swans is also valid for cyber-risk. Risk assessment is, as in the general case, mainly suited to capturing gray swans. So what is a gray swan in cyberspace? As we see it, many zero-day vulnerabilities are gray swans, for example. In the same way as economists do risk assessments taking a stock exchange crash into consideration, computer scientists do risk assessments addressing zero-day vulnerabilities. We know from experience that security relevant bugs may pop up even in mature software that has been carefully tested and has been in use for many years. Since cyber-systems are computerized to a large degree we have possibilities for testing, surveillance, and monitoring that may not easily be implemented in the general case, and that may make the detection of gray swans easier. On the other hand, the existence and influence of cyberspace has a complicating effect. In particular, cyberspace is highly dynamic. Hence, a valid threat picture today may not be valid tomorrow. To cope with the dynamics of cyberspace and therefore also of cyber-systems we may aim for more dynamic risk models whose risks, vulnerabilities, and threats are defined, measured, and as much as possible updated automatically in real time as functions of low-level indicators.

14.6 Further Reading

We have already recommended the influential work [77] of Nassim Nicholas Taleb on black swans as well as Gerd Gigerenzer's book [20] on natural frequencies. Regarding the comprehension of numbers there is also specialized literature available [65]. Contingency planning is a large subject on which many have published; see for example ISO 22301 on societal security [27].

For a classification of evolution in the context of risk assessment and a corresponding case study, see [46, 48]. Initial ideas for how to automatize or semi-automatize risk assessment to keep up with scaling and system evolution have been proposed by several authors [68, 42, 69]. The field is however immature.

Chapter 15
Conclusion

We have structured the conclusion into three parts. First we draw conclusions on the general theme of cyber-risk management as described in Parts I and II. Then we do the same for the four issues addressed in further detail in Part III. A technical brief is by its very definition short; hence, much has just been touched on and even more has not been covered at all. We end this chapter by identifying some of these issues.

15.1 What We Have Put Forward in General

Cyber-risk management is not fundamentally different from risk management in general; as we have explicated in the first two parts of this book, we recommend stakeholders to conduct cyber-risk management by following the processes and recommendations of established standards and practices on risk management.

There are however aspects of cyber-systems that make cyber-risk management challenging. The main feature in this respect is the use of cyberspace. Cyber-systems and cyberspace have brought significant improvements for individuals, businesses, and society as a whole within numerous areas, including social life, public services, trade and economy, entertainment, and critical infrastructures. At the same time, the use of and dependence on cyberspace has introduced a number of new threats and vulnerabilities.

In order to understand how to conduct cyber-risk management in an effective and efficient way it is necessary to understand the kinds of systems that we are concerned with, as well as the nature of the risks these systems are exposed to. This is why we have devoted separate chapters to cyber-systems, cybersecurity, and cyber-risk management in the first part of this book.

One important aspect of cyber-risk is the distinction between malicious cyber-risk and non-malicious cyber-risk. The distinction has implications for how we assess and handle cyber-risk, and we have therefore organized much of the contents in the two first parts of the book to account for this. The possibility of malicious threats requires a strong focus on human intent, motives, and capabilities. This has

© The Author(s) 2015
A. Refsdal et al., *Cyber-Risk Management*, SpringerBriefs
in Computer Science, DOI 10.1007/978-3-319-23570-7_15

led to the publication of dynamically evolving catalogues and repositories documenting potential cyber-threats, exploits, and vulnerabilities to malicious attacks, as well as techniques for the modeling of malicious threats. At the same time, the many possibilities of accidental and unintended incidents require a similar focus on non-malicious threats, including both the technical and the sociotechnical aspects of cyber-systems. Together with the wide extension of cyberspace, and therefore the wide possibilities for threats to arise, the different ways in which to tackle malicious and non-malicious threats represent a challenge for cyber-risk management. This challenge must be handled in a methodical manner, as we have put forward in this book.

Another major challenge regarding cyber-risk management is that cyberspace evolves rapidly and often in a manner that is difficult to predict. Cyber-systems must be able to cope with this evolution. In fact, cyber-systems are forced to evolve in response to the evolution of cyberspace. This requires increased focus on monitoring and risk assessment in real time as part of the overall cyber-risk management.

Although cyber-systems are challenging from a risk management point of view, there are also features of cyber-systems that we can take advantage of and that have a simplifying effect. The fact that cyber-systems are computerized to a large degree is beneficial when it comes to data collection, which is why we have stressed the use of techniques such as monitoring and testing throughout Part I and Part II of this book. Moreover, computerized harvesting of data may reduce uncertainty in risk assessment. In fact, the possibility of data collection in real time provides a foundation for real-time risk assessment.

15.2 What We Have Put Forward in Particular

Risk level may be measured in multiple ways. We have presented the two-factor approach based on consequence and likelihood, which is the one most commonly used in practice. We have also considered an alternative approach employing three factors developed for the security domain, and we have discussed the use of more than three factors. Which approach to use and how to use it depends on the context and your risk assessment situation. Which data are available is an important parameter when deciding how to measure risk level. If you have good data on frequency and consequence, and not on other factors, you will probably go for the two-factor approach, and accordingly for other measures if they are favored by the data available.

Estimating or measuring likelihood tends to be difficult. One reason is that there may be considerable uncertainty as to what the likelihood is. Another reason is that in some cases there is a lack of experience or historical data with respect to the event in question. A third reason is that we may all too easily complicate the task ourselves by selecting quantitative likelihood scales that are badly suited to the task. In general, we do not recommend using probabilities when interacting with human beings in a risk assessment situation. A probability is always defined implicitly with respect to some interval or context, and the existence of this implicit interval or context is easily overlooked or misunderstood leading to bad estimates and confusion. In most cases natural frequencies are better suited to risk analysis purposes.

Make sure not to confuse likelihood with uncertainty. It makes good sense to document uncertainty for each risk factor separately. When working quantitatively, in our experience a practical approach to take uncertainty into consideration is to use intervals. When employing qualitative scales, uncertainty may be characterized separately, for example, as a separate natural language expression for each measurement. In a risk assessment, whether the level of uncertainty is tolerable depends on to what extent the uncertainty impacts the decision procedure. If it does not, the uncertainty is at an acceptable level.

Risk assessment has its limitations. In particular, as we emphasized in Chap. 14, risk assessment will in most cases be of little help in identifying and predicting black swans. On the other hand, we have argued that risk assessment may be well suited to coping with gray swans. To reduce the chance of gray swans not being considered we have argued that it is often fruitful to observe the target of assessment from different viewpoints. This has several implications for how we conduct risk identification. For example, we should involve all relevant stakeholder roles, split the risk identification into a set of independent processes, and embed the use of different kinds of tools in the risk identification process.

15.3 What We Have not Covered

The main focus of this book is on cyber-risk assessment. The more general and continuous risk management activities corresponding to the processes for "communication and consultation" as well as "monitoring and review" are just covered briefly.

Another topic that we have touched upon, but which requires much more careful consideration, is system evolution and its implications for risk assessment. Real-time risk assessment is another important aspect of risk management not covered by this book. We believe risk assessment in real time will become more and more important in order to cope with ever more dynamic cyber-systems.

Within the general fields of cyber-systems and cybersecurity there are numerous sub-fields imposing more specialized challenges to risk management. Privacy is one such sub-field; compliance, cloud computing, and big data are other examples, none of which are covered by this book.

Glossary

Absolute scale A *ratio scale* in which one value among the values may be understood as the smallest unit.

Aleatory uncertainty Uncertainty due to inherent randomness that cannot be removed without redesigning the object at hand.

Asset Anything of value to a *party*.

Attack surface All of the different points where an attacker or other *threat source* could get into the *cyber-system*, and where information or data can get out.

Black swan An *incident* that is extremely rare and unexpected, but has very significant *consequences*.

Communication and consultation Activities aiming to provide, share, or obtain information and to interact with *stakeholders* regarding the *management of risk*.

Consequence The impact of an *incident* on an *asset* in terms of harm or reduced asset value.

Context establishment Activities aiming to specify the *external* and *internal context* as well as providing all the input that is needed for the following steps of *risk assessment*.

Critical infrastructure protection Prevention of the disruption, disabling, destruction, or malicious control of critical infrastructure.

Cyber-physical system A *cyber-system* that controls and responds to physical entities through actuators and sensors.

Cyber-risk A *risk* that is caused by a *cyber-threat*.

Cybersecurity The protection of *cyber-systems* against *cyber-threats*.

Cyberspace A collection of interconnected computerized networks, including services, computer systems, embedded processors, and controllers, as well as information in storage or transit.

Cyber-system A *system* that makes use of a *cyberspace*.

Cyber-threat A *threat* that exploits a *cyberspace*.

© The Author(s) 2015
A. Refsdal et al., *Cyber-Risk Management*, SpringerBriefs
in Computer Science, DOI 10.1007/978-3-319-23570-7

Difference scale An *ordinal scale* such that equality of difference at the level of values implies that the corresponding phenomena are equally distinct.

Epistemic uncertainty Uncertainty due to ignorance or lack of evidence.

External context Includes the relationship with external *stakeholders* as well as the relevant societal, legal, regulatory, and financial environment.

Focus of assessment The main area or central area of attention in a *risk assessment*.

Frequency A measure of the number of occurrences of something per unit of time.

Gray swan An *incident* which has far-reaching *consequences*, but, unlike a *black swan*, can be anticipated to a certain degree

Incident An event that harms or reduces the value of an *asset*.

Information security Preservation of confidentiality, integrity, and availability of information.

Internal context Includes the relevant goals, objectives, policies, and capabilities that may determine how *risk* should be *assessed*.

Likelihood The chance of something to occur.

Malicious cyber-risk A *cyber-risk* that is (at least partly) caused by a malicious *threat*.

Monitoring Continual checking, supervising, critically observing, or determining the current *risk* status in order to identify deviation from the expected or required *risk* status.

Nominal scale A scale that allows us to categorize the phenomena we are concerned with into disjoint categories.

Non-malicious cyber-risk A *cyber-risk* that is caused by a non-malicious *threat*.

Ordinal scale A *nominal scale* whose categories or values are ordered.

Party An organization, company, person, group, or other body on whose behalf a *risk assessment* is conducted.

Probability A measure of the chance of something to occur expressed as a number between 0 and 1.

Qualitative scale A *nominal scale* or an *ordinal scale*.

Quantitative scale A *difference scale* or a *ratio scale*.

Ratio scale A *difference scale* such that equality of ratios at the level of values implies that the corresponding phenomena are equally distinct.

Review Activities aiming to determine the suitability, adequacy, and effectiveness of the *risk management* process and framework, as well as *risks* and *treatments*.

Risk The *likelihood* of an *incident* and its *consequence* for an *asset*.

Risk analysis Activities aiming to estimate and determine the *risk level* for identified *risks*.

Risk assessment Activities aiming to understand and document the risk picture for specific parts or aspects of a *system* or an organization.

Risk assessment process A five-step process for *risk assessment* consisting of *context establishment*, *risk identification*, *risk analysis*, *risk evaluation*, and *risk treatment*.

Risk evaluation Activities involving the comparison of the *risk analysis* results with the *risk evaluation criteria* to determine which *risks* should be considered for *treatment*.

Risk evaluation criteria The terms of relevance by which the significance of *risk* is *assessed*.

Risk identification Activities aiming to identify, describe, and document *risks* and possible causes of *risk*.

Risk level The magnitude of a *risk* as derived from its *likelihood* and *consequence*.

Risk management Coordinated activities to direct and control an organization with regard to *risk*.

Risk model Any representation of *risk* information, such as *threats*, *vulnerabilities*, *incidents*, and how they are related.

Risk treatment Activities aiming to identify and select means for *risk* mitigation and reduction.

Safety Protection of life and health by the prevention of physical injury caused by damage to property or to the environment.

Scope of assessment The extent or range of a *risk assessment*.

Smart grid An electricity distribution network that can monitor the flow of electricity within itself and automatically adjust to changing conditions.

Stakeholder of risk assessment Any person or organization that may affect or be affected by the subject of the *risk assessment*.

Stakeholder of risk management Any person or organization that may affect or be affected by the organization that is the subject of the *risk management*.

System A set of related entities that forms an integrated whole and has a boundary to its surroundings.

Target of assessment The parts and aspects of the *system* that is the subject of the *risk assessment*.

Threat An action or event that is caused by a *threat source* and that may lead to an *incident*.

Threat source The potential cause of an *incident*.

Treatment An appropriate measure to reduce *risk level*.

Vulnerability A weakness, flaw, or deficiency that can be exploited by a *threat* to cause harm to an *asset*.

References

1. Avižienis, A., Laprie, J.C., Randell, B., Landwehr, C.: Basic concepts and taxonomy of dependable and secure computing. IEEE Transactions on Dependable and Secure Computing **1**, 11–33 (2004)
2. Bajpai, S., Sachdeva, A., Gupta, J.P.: Security risk assessment: Applying the concepts of fuzzy logic. Journal of Hazardous Materials **173**, 258–264 (2010)
3. Beckers, K.: Pattern and security requirements – Engineering-based establishment of security standards. Springer (2015)
4. Beckers, K., Heisel, M., Solhaug, B., Stølen, K.: ISMS-CORAS: A structured method for establishing an ISO 27001 compliant information security management system. In: Engineering Secure Future Internet Services, *Lecture Notes in Computer Science*, vol. 8431, pp. 315–344. Springer (2014)
5. Ben-Gal, I.: Bayesian networks. In: Encyclopedia of Statistics in Quality and Reliability. Wiley (2007)
6. Böhme, R., Schwartz, G.: Modeling cyber-insurance: Towards a unifying framework. In: Workshop on the Economics of Information Security (WEIS'10) (2010)
7. CNSS: Instruction No. 4009 – Committee on National Security Systems (CNSS) Glossary (2015)
8. Common Criteria: Common methodology for information technology security evaluation – Evaluation methodology, v3.1, rev. 4 (2012)
9. Encyclopedia Britannica: Venn diagram. Online: http://www.britannica.com/EB-checked/topic/625448/Venn-diagram [Accessed April 17, 2015]
10. Erdogan, G., Li, Y., Runde, R.K., Seehusen, F., Stølen, K.: Approaches for the combined use of risk analysis and testing: A systematic literature review. Journal on Software Tools for Technology Transfer **16**, 627–642 (2014)
11. Erdogan, G., Seehusen, F., Stølen, K., Hofstad, J., Aagedal, J.Ø.: Assessing the usefulness of testing for validating and correcting security risk models based on two industrial case studies. International Journal of Secure Software Engineering **6**, 90–112 (2015)
12. European Commission: COM(2011) 163 final – On critical information infrastructure protection – Achievements and next steps: Towards global cyber-security (2011)
13. European Commission: JOIN(2013) 1 final – Cybersecurity strategy of the European Union – An open, safe and secure cyberspace (2013)
14. European Network and Information Security Agency: Deliverable-2011-12-09 – Protecting industrial control systems – Recommendations for Europe and member states. (2011)
15. European Network and Information Security Agency: Incentives and barriers of the cyber insurance market in Europe (2012)
16. EUROPOL: The Internet organised crime threat assessment (iOCTA) (2014)
17. Ferson, S., Ginzburg, L.R.: Different methods are needed to propagate ignorance and variability. Reliability Engineering and System Safety **54**, 133–144 (1996)

© The Author(s) 2015
A. Refsdal et al., *Cyber-Risk Management*, SpringerBriefs
in Computer Science, DOI 10.1007/978-3-319-23570-7

18. Galtung, J.: Theories and methods of social research, revised edn. Universitetsforlaget (1969)
19. Geisberger, E., Broy, M. (eds.): Integrierte Forschungsagenda Cyber-Physical Systems. Springer (2012)
20. Gigerenzer, G.: Calculated risks – How to know when numbers deceive you. Simon & Schuster (2002)
21. Grossmann, J., Schneider, M., Viehmann, J., Wendland, M.F.: Combining risk analysis and security testing. In: 6th International Symposium on Leveraging Applications of Formal Methods, Verification and Validation (ISoLA'14), *Lecture Notes in Computer Science*, vol. 8803, pp. 322–336. Springer (2014)
22. Hajek, P.: Fuzzy logic. In: The Stanford Encyclopedia of Philosophy, fall edn. Stanford University (2010)
23. International Electrotechnical Commission: IEC/TR 61508-0 – Functional safety of electrical/electronic/programmable electronic safety-related systems – Part 0: Functional safety and IEC 61508 (2005)
24. International Electrotechnical Commission: IEC 62502 – Analysis techniques for dependability – Event tree analysis (ETA) (2010)
25. International Organization for Standardization: ISO 31000 – Risk management – Principles and guidelines (2009)
26. International Organization for Standardization: ISO Guide 73 – Risk management – Vocabulary (2009)
27. International Organization for Standardization: ISO 22301 – Societal security – Business continuity management systems – Requirements (2012)
28. International Organization for Standardization / International Electrotechnical Commission: ISO/IEC 27032 – Information technology – Security techniques – Guidelines for cybersecurity (2005)
29. International Organization for Standardization / International Electrotechnical Commission: ISO/IEC 27004 – Information technology – Security techniques – Information security management – Measurement (2009)
30. International Organization for Standardization / International Electrotechnical Commission: ISO/IEC 31010 – Risk management – Risk assessment techniques (2009)
31. International Organization for Standardization / International Electrotechnical Commission: ISO/IEC 25010 – Systems and software engineering – Systems and software quality requirements and evaluation (SQuaRE) – System and software quality models (2011)
32. International Organization for Standardization / International Electrotechnical Commission: ISO/IEC 27005 – Information technology – Security techniques – Information security risk management (2011)
33. International Organization for Standardization / International Electrotechnical Commission: ISO/IEC 27001 – Information technology – Security techniques – Information security management systems – Requirements (2013)
34. International Organization for Standardization / International Electrotechnical Commission: ISO/IEC 27002 – Information technology – Security techniques – Code of practice for information security controls (2013)
35. International Organization for Standardization / International Electrotechnical Commission: ISO/IEC 27000 – Information technology – Security techniques – Information security management systems – Overview and vocabulary (2014)
36. International Telecommunication Union: ITU-T X.1055 – Risk management and risk profile guidelines for telecommunication organizations (2008)
37. International Telecommunication Union: ITU-T X.1205 – Data networks, open system communications and security – Telecommunication security – Overview of cybersecurity (2008)
38. ISACA: COBIT 5: A business framework for the governance and management of enterprise IT (2012)
39. ISACA: COBIT for risk (2013)
40. Jensen, F.V., Nielsen, T.D.: Bayesian networks and decision graphs. Springer (2007)
41. Jøsang, A.: A logic for uncertain probabilities. International Journal of Uncertainty, Fuzziness and Knowledge-Based Systems **9**, 279–311 (2001)

42. Kaliski, B., Pauley, W.: Toward risk assessment as a service in cloud environments. In: 2nd USENIX Conference on Hot Topics in Cloud Computing. USENIX Association (2010)
43. Kurzenhäuser, S.: Natural frequencies in medical risk communication: Applications of a simple mental tool to improve statistical thinking in physicians and patients. Ph.D. thesis, Freie Universität Berlin (2003)
44. Lee, E.: Cyber physical systems: Design challenges. Tech. Rep. UCB/EECS-2008-8, University of California at Berkeley (2008)
45. Line, M.B., Johansen, G., Sæle, H.: Risikovurdering av AMS. Tech. Rep. A22318, SINTEF (2012)
46. Lund, M.S., Solhaug, B., Stølen, K.: Evolution in relation to risk and trust management. Computer **43**, 49–55 (2010)
47. Lund, M.S., Solhaug, B., Stølen, K.: Model-driven risk analysis – The CORAS approach. Springer (2011)
48. Lund, M.S., Solhaug, B., Stølen, K.: Risk analysis of changing and evolving systems using CORAS. In: 11th International School on Foundations of Security Analysis and Design (FOSAD'11), *Lecture Notes in Computer Science*, vol. 6858, pp. 231–274. Springer (2011)
49. Macmillan Dictionary: Black swan. Online: http://www.macmillandictionary.com/dictionary/british/black-swan [Accessed March 9, 2015]
50. Macmillan Dictionary: Grey swan. Online: http://www.macmillandictionary.com/dictionary/british/grey-swan [Accessed March 9, 2015]
51. MITRE: Common attack pattern enumeration and classification (CAPEC). Online: https://capec.mitre.org/ [Accessed December 8, 2014]
52. MITRE: Common weakness enumeration (CWE). Online: http://cwe.mitre.org/ [Accessed December 8, 2014]
53. National Institute of Standards and Technology: NISTIR 7628 – Guidelines for smart grid cyber security – Vol. 3: Supportive analyses and references (2010)
54. National Institute of Standards and Technology: Guide for conducting risk assessments, special publ. 800-30 (2012)
55. National Institute of Standards and Technology: Framework for improving critical infrastructure cybersecurity, v1.0 (2014)
56. National Institute of Standards and Technology: NISTIR 7628 – Guidelines for smart grid cyber security – Vol. 1: Smart grid cyber security strategy, architecture, and high-level requirements, rev. 1 (2014)
57. National Security Telecommunications and Information Security Committee: National information assurance certification and accreditation process (NIACAP) (2000)
58. Object Management Group: OMG Unified Modeling Language (OMG UML) – Superstructure, OMG Document:formal/2009-02-02, v2.2 (2009)
59. O'Hagan, T.: Dicing with the unknown. Significance **1**, 132–133 (2004)
60. OWASP: Attack surface analysis cheat sheet. Online: http://www.owasp.org/index.php/Attack_Surface_Analysis_Cheat_Sheet [Accessed December 5, 2014]
61. OWASP: The open web application security project. Online: http://www.owasp.org [Accessed December 5, 2014]
62. OWASP: Risk rating methodology. Online: http://www.owasp.org/index.php/OWASP_Risk_Rating_Methodology [Accessed March 11, 2014]
63. OWASP: OWASP top 10 – The ten most critical web application security risks (2013)
64. OWASP: Testing guide, v4.0 (2013)
65. ProClarity: Effectively communicating numbers – Selecting the best means and manner of display (2005)
66. PwC: The global state of information security survey (2015)
67. Rajkumar, R., Lee, I., Sha, L., Stankovic, J.: Cyber-physical systems: The next computing revolution. In: 47th Design Automation Conference (DAC'10), pp. 731–736. ACM (2010)
68. Refsdal, A., Stølen, K.: Employing key indicators to provide a dynamic risk picture with a notion of confidence. In: 3rd IFIP WG 11.11 International Conference on Trust Management (IFIPTM'09), pp. 215–233. Springer (2009)

69. Saripalli, P., Walters, B.: QUIRC: A quantitative impact and risk assessment framework for cloud security. In: 3rd International Conference on Cloud Computing (CLOUD'10), pp. 280–288. IEEE Computer Society (2010)
70. Schneier, B.: Attack trees: Modeling security threats. Dr. Dobb's Journal **24**, 21–29 (1999)
71. Sentz, K., Ferson, S.: Combination of evidence in Dempster-Shafer theory. Tech. Rep. SAND2002-0835, Sandia National Laboratories (2002)
72. Solhaug, B., Stølen, K.: Uncertainty, subjectivity, trust and risk – How it all fits together. In: 7th International Workshop on Security and Trust Management, *Lecture Notes in Computer Science*, vol. 7170, pp. 1–5. Springer (2012)
73. Sophos: Security threat report (2014)
74. Stevens, S.S.: On the theory of scales of measurement. Science **103**, 677–680 (1946)
75. Swiderski, F., Snyder, W.: Threat modeling. Microsoft Press (2004)
76. Symantec: Internet security threat report (2014)
77. Taleb, N.T.: The black swan: The impact of the highly improbable. Brockman (2007)
78. Tanenbaum, A.S.: Computer networks, 4 edn. Prentice Hall (2003)
79. Tøndel, I.A., Jaatun, M.G., Line, M.B.: Security threats in Demo Steinkjer. Tech. Rep. A23351, SINTEF (2012)
80. United States Department of Homeland Security: NIPP 2013 - Partnering for critical infrastructure security and resilience (2013)
81. United States Government Accountability Office: Critical infrastructure protection – Department of Homeland Security faces challenges in fulfilling cybersecurity responsibilities (2005)
82. Verizon: Data breach investigations report (2014)
83. The White House, U.S.: International strategy for cyberspace – Prosperity, security and openness in a networked world (2011)
84. Wikipedia: Willem de Vlamingh. Online: http://en.wikipedia.org/wiki/Willem_de_Vlamingh [Accessed March 9, 2015]
85. Willis, H.H., Morral, A.R., Kelly, T.K., Medby, J.J.: Estimating terrorism risk. Tech. Rep. MG-388-RC, RAND Corporation (2005)
86. Wolf, W.: Cyber-physical systems. Computer **42**, 88–89 (2009)
87. World Economic Forum: Global risks, 10 edn. (2015)

Index

access control, 22, 42
acronym
  list of, xi, 3
adversary, 35–40, 64
AGRA, vi
Aida Omerovic, vi
al Qaeda, 123
Alexander Pretschner, vi
Algirdas Avižienis, 32
AMI, *xi*, 4, 51, 66
Andrew Stuart Tanenbaum, 26
Aristotelis Tzafalias, vi
ARPANET, *xi*, 25
Aslak Wegner Eide, vi
asset, 4, **10**, 29, 35, 57, 61, 81, 93, 99, 107,
    112, 125, *133*
  identification, 17, 57
  information, 30, 41
  infrastructure, 30
  intangible, 6, 41
  value, 11, 107
asset-driven
  process, 40
assumption, 16, 57
attack, v, 1, 32, 38, 55, 61, 82, 98, 110, 123,
    130
  DoS, 29, 33
  injection, 29, 70
  likelihood, 47
  malicious, v, 110, 130
  opportunity, 43, 82, 110
  pattern, 47, 70
  strategy, 37
  surface, **37**, 55, 61, 121, *133*
  virus, 33
attack tree, 19, 20
attacker, 35–40, 55, 64, 110, 125

group, 109
  knowledge, 43, 89, 110
  modeling, 47
  motive, 43, 110
  skill, 43, 67, 82, 110, 116

Barack Obama, 1
Bayesian network, 19, 20, 121
bibliography, 3
big data, 131
brainstorming, 19, 39, 45, 63

CAPEC, *xi*, 39, 43, 47, 67, 89
catalogue, 39, 89, 130
central system, 54, 62, 82, 96, 98
checklist, 19
choke, 54, 68, 83, 96, 98
Christian W. Probst, vi
CIIP, *xi*, 30, 32
CIP, *xi*, 30–32
cloud computing, 42, 45, 131
CNSS, *xi*, 32
COBIT, *xi*, 6
Common Criteria, 43, 47, 110, 116
communication and consultation, **13**, 34, 131,
    133
compliance, 15, 37, 52, 103, 131
CONCERTO, vi
consequence, 4, **11**, 31, 35, 52, 62, 81, 91, 101,
    107, 111, 117, 123–127, 130, *133*
  estimation, 20, 43
  factor, 5, 107–110, 113, 130
  scale, 17, 58, 88, 115
  value, 17, 58, 94, 112
context
  description, 16
  establishment, 4, **16**, 37, 51–61, 81, 92, *133*

© The Author(s) 2015
A. Refsdal et al., *Cyber-Risk Management*, SpringerBriefs
in Computer Science, DOI 10.1007/978-3-319-23570-7

Printed in the United States
By Bookmasters